50个最好的别墅

50 Top House

佳图文化 编

中国林业出版社

图书在版编目（CIP）数据

50个最好的别墅 / 王志 主编. -- 北京 : 中国林业出版社, 2011.3

ISBN 978-7-5038-6066-9

Ⅰ. ①5… Ⅱ. ①王… Ⅲ. ①别墅－室内设计：空间设计 Ⅳ. ①TU241.1

中国版本图书馆CIP数据核字(2011)第006303号

策划：佳图文化
主编：王志　龙志伟

中国林业出版社·环境景观与园林园艺图书出版中心
责任编辑：李 顺
电话、传真：（010）83229512

出 版：中国林业出版社（100009 北京西城区德内大街刘海胡同7号）
印 刷：上海瑞时印刷有限公司
发 行：新华书店北京发行所
电 话：（010）83224477
版 次：2011年3月第1版
印 次：2011年3月第1次
开 本：889mm × 1092mm 1 / 12
印 张：26
字 数：500千字
定 价：228.00元

Preface 前言

"50 Top Houses" has carefully selected 50 most beautiful villas from a lot of latest domestic and overseas projects with excellent architectural and interior designs. The boutique villa projects of Germany, Netherlands, Greece, Italy, UK, USA, Australia, Japan, Croatia, Argentina and other places are the latest representative works of the internationally renowned designers'. These villas are designed in different styles with unique characteristics to be the top-level works. All the projects have live-action pictures, plan drawings and elevation drawings. Each project is carefully analyzed from architectural design, interior design and landscape design to show its design idea and uniqueness. It is worth reading and collecting by the architectural designers, interior designers, real estate developers, the related students and owners.

《50个最好的别墅》从大量最新国内外别墅案例中精挑出50个最美的建筑及空间设计案例。这些来自德国、荷兰、希腊、意大利、英国、美国、澳大利亚、日本、克罗地亚、阿根廷等世界各地不同地区的精美别墅设计案例，都是国际知名设计师的最新代表作品，造型各异、极具创意特色，堪称当下别墅建筑与空间设计的顶级之作。全部案例以专业拍摄的实景图片为主，配备齐全的平面图、立面图，从建筑设计、室内设计、景观设计几个角度详细剖析这些成功案例的设计灵感与独特之处，非常值得建筑设计师、室内设计师、开发商、相关专业学生及业主等读者收藏赏鉴。

CONTENTS 目录

King Residence

国王金居

Floor area: 399.5 m²

建筑面积：399.5m²

Architect: John Friedman Alice Kimm Architects

建筑：约翰・弗瑞德曼・爱丽思・金姆建筑师事务所

The King Residence is a 4,300 square foot single family residence located in the Sunset Park section of Santa Monica. Situated on a gently sloping, wedge-shaped corner lot, the design rejects the standard public front yard/private backyard typology, opting instead for a structure whose living spaces and bedrooms open onto a relatively large exterior patio/garden/entryway that faces the public streets and surrounding houses. While this arrangement puts much of the family's communal and individual life on display, this is a positive feature for the Kings.

By responding to the street corner with a void, instead of a volume, and setting the house towards the rear of the site, the house does not crowd the neighborhood – it exudes a sense of politeness, of generosity. This quality is reinforced by its composition of overlapping planes and layers, which break up its mass and project a sense of permeability; from down the street, one can even see straight through the house. These intentions carry through to the strategies for cladding the house, the best example being the way in which the two-toned cement board pattern, echoing the character of leaves on nearby trees, helps to dematerialize the structure. The wood screens have a similar effect.

Both inside and outside, the house's design manages to be both serious as well as casual – to create the conditions for a flexible, open, communally-oriented lifestyle, while at the same time employing a refined palette of materials and details. The generous use of natural light activates the house and creates excellent lighting for the family's art collection. A number of sustainable features are also incorporated. For the Kings, the house functions perfectly for their individual and collective lifestyles, and embodies the values in which they believe.

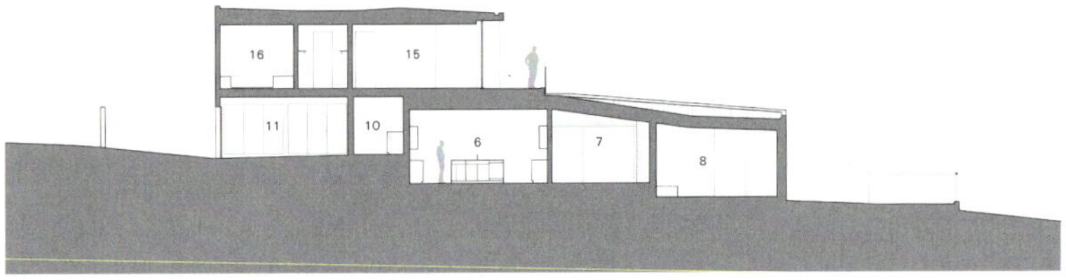

BUILDING SECTION A

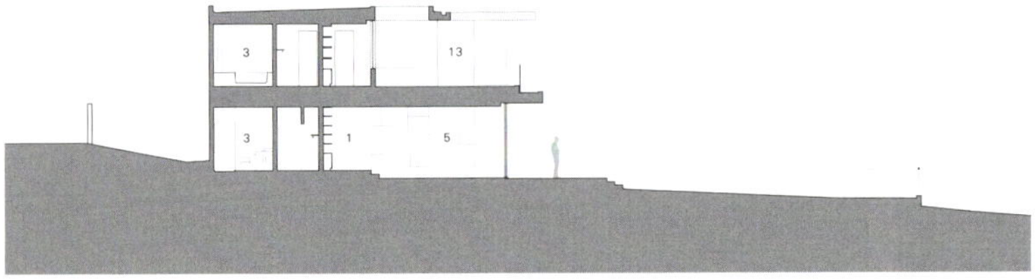

BUILDING SECTION B

LEGEND

1 FOYER
2 OFFICE
3 BATHROOM
4 WINE CELLAR
5 LIVING ROOM
6 KITCHEN
7 DINING ROOM
8 HANG OUT ROOM
9 POWDER ROOM
10 LAUNDRY ROOM
11 GARAGE
12 FOUNTAIN
13 PATIO
14 BEDROOM
15 MASTER BEDROOM
16 MASTER BATHROOM

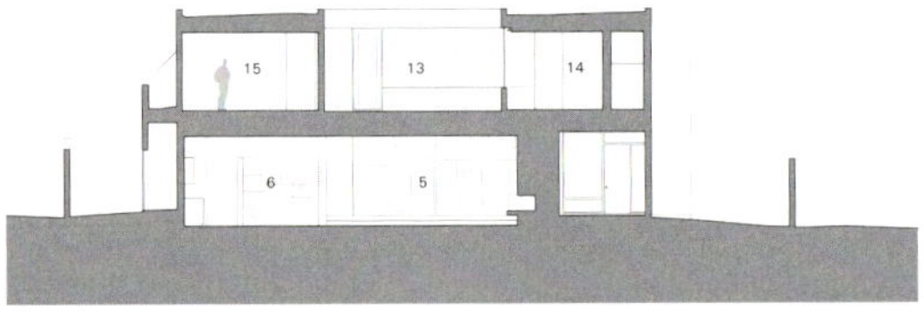

BUILDING SECTION C

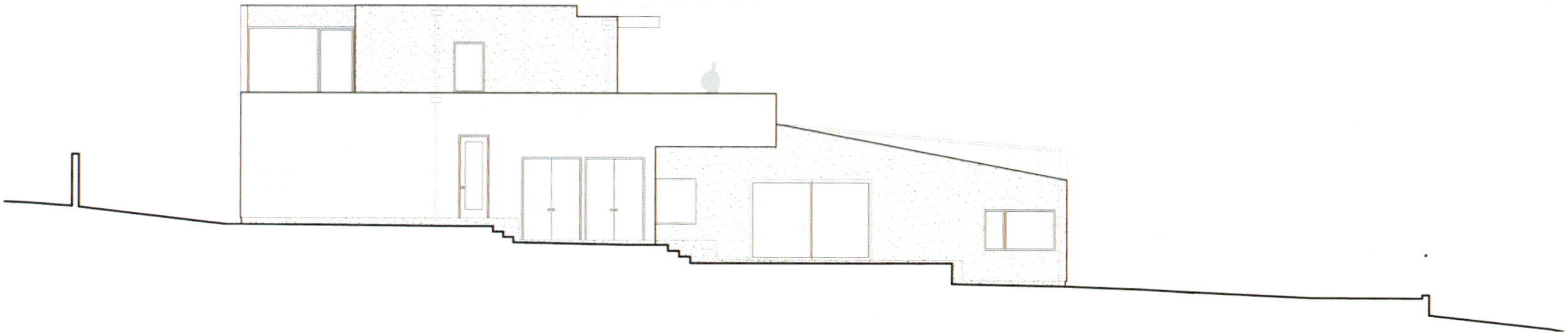

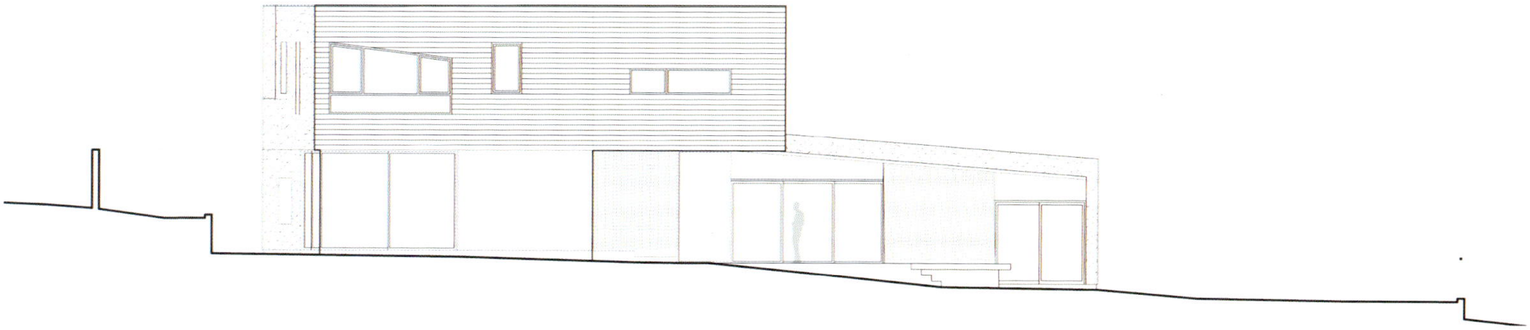

国王金居是一个4,300平方英尺（约 399.5m²）的独户住宅，位于圣塔莫尼卡市的日落公园。处在缓坡之上，“V”型的转角，设计不落俗套，没有常见的公共前院（私人后院），而是采用一种独特的结构，客厅和卧室通向一个相对较大的外部天井（花园、入口通道），朝向大街和周围的住宅。而这种布局将家人的公共以及私人生活公布于众，这是国王金居的一大特色。

在街角处，留出空间，并且将住宅朝向本案的后面，住宅就不会使街坊显得拥挤了——它散发出礼貌优雅的迷人气息。重叠的平面和层次的融合更加突出了这一品质，它打破了厚重感，而映射出一种渗透感；从街边，可以看到这座住宅内部。这些用意贯穿整个住宅外饰策略，最好的例证即是两种色调的水泥板的装饰模式，与街边的枝繁叶茂相互呼应，而掩盖了房屋的结构。木屏风具有同样的效果。

住宅的内外设计都尽量既严肃又随意——为了营造出灵活、开放，以共同生活为主导的生活氛围，同时又利用精致的材料和优雅的细节来进行调色。充分的自然光照使得住宅活泼生动，为家庭艺术品收藏创造了极佳的光源。它也融入了一些可持续的元素。对于国王来说，住宅功能齐全，能满足主人个人和集体生活的需要，并且体现了主人的价值观念。

LATER

Northaven Residence

诺特天堂住宅

Designed by: Lionel Morrison FAIA

设计：美国建筑师协会会员莱昂内尔·莫里森

The genesis of the design was a very old and large oak tree found on the site. This tree together with the existing site contours, that seemed to spread from the tree like ripples in a pond, gave rise to the curved form of the house. The tree marks the point from which the arc springs. The natural geometric result of this is that all openings within the arc focus back to the tree.

Initially the house looks like an arced vessel contained within an exoskeleton of structural steel that in turn supported trellises and sunscreens. We used computer modeling extensively to assist in the design of the various screening devices. Additional shading is provided within the house via computer-controlled shades.

The house can be seen as a courtyard house albeit a curved one. Since the house is glazed almost entirely on the ground level the perception of the space runs from the lower garden wall through the house across the water to the tall limestone wall at the rear. The width of the water matches that of enclosed space.

Water is a critical and fundamental component of the house. As such it defines the horizontal plane of the courtyard and is given equal importance to the other architectural elements that form the outdoor space.

该设计源于这块场地上一棵非常古老的大橡树。以这棵橡树为中心，这个场地的轮廓像池塘的涟漪一样，这棵橡树与这个地方的轮廓增加了房子的曲线美。橡树是弧形伸展的起点，这种自然几何使弧形的每个开口又以橡树为中心。

一眼望去，房子看起来像一个以结构为外骨骼的弧形船，弧形船反过来又增强了钢结构和防晒作用。我们广泛使用电脑模拟辅助各种防日晒功能的设计。通过计算机控制光照，房子内有了更多的阴凉。

虽然是弧形的，房子可看作四合院。地面部分的房子几乎全是玻璃的，空间感从房子较低的园墙横跨水面一直延伸到后面的高大石灰岩壁。宽阔的水面与封闭的空间，相映成趣。

水是本方案的一个关键和基本组成部分。水确定了庭院的水平面，与形成户外空间的其他建筑元素相比，水得到了同样的重视。

SITE PLAN
0 20 40

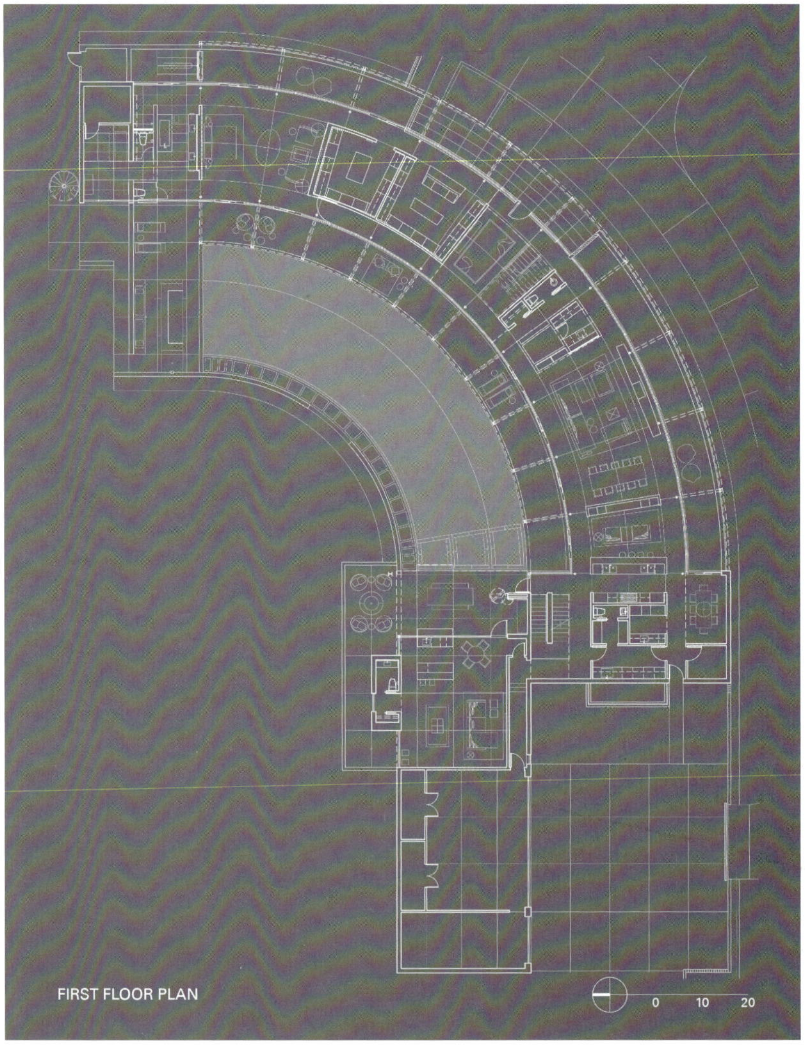
FIRST FLOOR PLAN
0 10 20

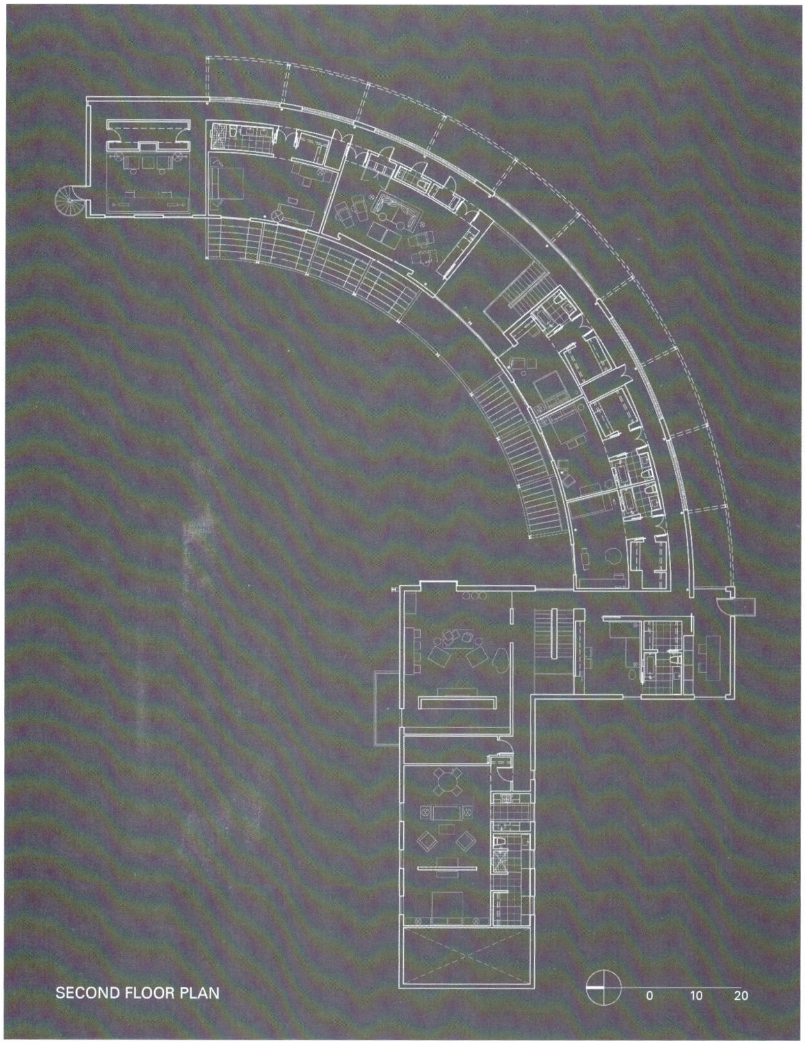
SECOND FLOOR PLAN
0 10 20

EDDI's House

艾迪屋

Designed by: Edward Suzuki Associates
Architect: Edward Suzuki

设计：爱德华 · 铃木联合公司
建筑：爱德华 · 铃木

This is the first case of an industrialized house in Japan (and perhaps in the world). The house is named after the architect's nickname and also stands for Edward Daiwa-House Design Innovation. Pre-designed and pre-fabricated and sold as a ready-made "product," the buyers' characteristics, lifestyles, site conditions, etc. are not known at the time of design.

The design concept is "Go in to go out," meaning that the house has an outside patio at the center of the house onto which each and every room looks out. Furthermore, a balcony is placed above and adjacent to the central patio overlooking it such that this combination of open spaces creates an "interface" between the outside proper and the inside that acts a filter, a buffer, or a cushion between the two zones. In this way the design is such that the house does not need to depend on the contextual conditions in which it is to be situated. No matter how unfavorable the surroundings may be, the house would function comfortably within the mini-cosmos of its own boundaries. Should the outside conditions be favorable for a change, all that a buyer needs to do is to request extra openings onto the outside in order to bring the assets of the outside into the house.

The house with its central patio is designed to allow as much of not only direct natural sunlight but also cross ventilation as possible to provide for maximum energy efficiency. EDDI's House now boasts six basic types of plans with sixty-nine variations to accommodate as much of the varying site conditions and characteristics.

1F plan

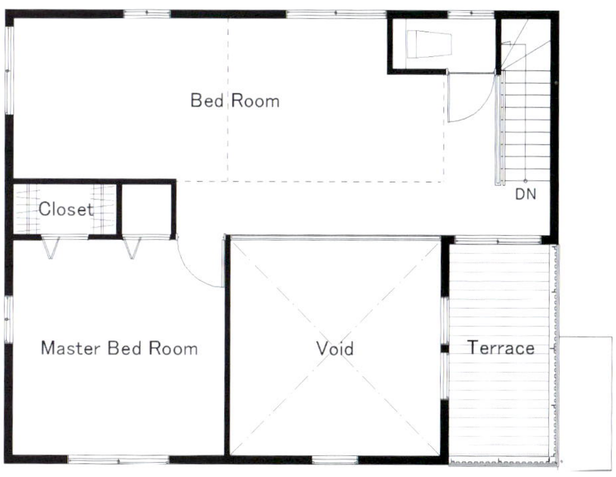

2F plan

这是日本（或许是世界上）首座工业化住宅。这座住宅以建筑师的昵称命名，也代表着爱德华大和——住宅设计的创新。这是一座事先设计并建造，然后作为完工“产品”进行销售的住宅，所以在进行设计时，“买主”的性格、生活方式、对选址的要求等等都是未知的。

设计理念是“由进而出”， 也就是说这座住宅在中心有一个外部天井，每个房间都面朝天井。而且，在天井之上并与其毗连，有一个阳台俯瞰着天井，这种开放空间的融合，刚好是分隔外部和内部的一种“界面”，是两者之间的一种过滤、缓冲或者说缓和。这样，住宅的设计就不需要依它所在的周围环境条件而定。无论周围环境多么不利，有自己的小环境，这座住宅都会非常舒适。如果室外条件允许，买主要改变一下风格的话，那么所需要做的就是另外打开几处出口把室外的宝贵资源引入室内。

住宅的中心天井不仅旨在尽可能使自然的阳光直射进来，而且也能对流通风，实现最好的节能效果。值得一提的是，“艾迪屋”目前有6种布局方案和69种可变元素，以尽可能满足买主的各种要求，贴合他们不同的性格。

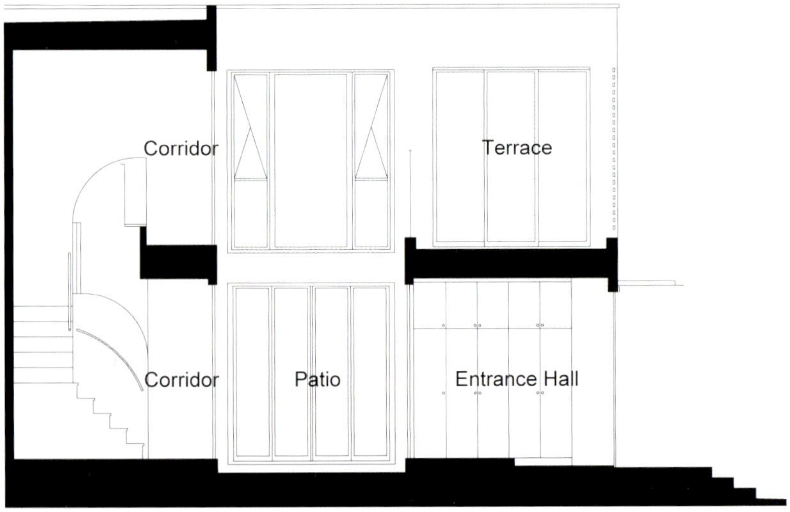

A-A' SECTION

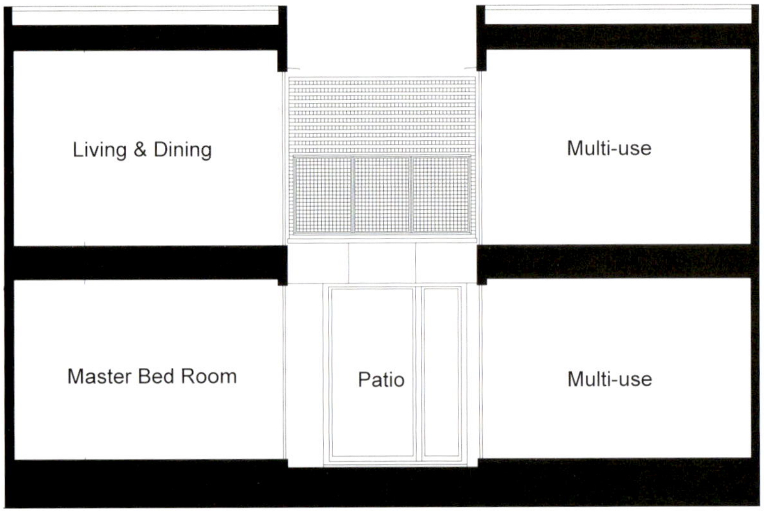

B-B' SECTION

Bed Room
S
U
Closet
Bed Room
Entrance Hall
Master Bed Room

R
Storage
Kitchen
D
Corridor
Void
Living & Dining
Multi-use
Terrace

Berkshire Residence

波克夏居

Floor area: 510.97 m^2 Architecture designed by: Morrison Seifert Murphy

建筑面积：510.97m^2 建筑设计：莫里森・塞弗特・墨菲

This site is in a traditional single-family neighborhood with large mature trees in front and a busy thoroughfare in the rear. These opportunities and limitations became the genesis of the design of the house.

The plan of the house is a U-shape that turns its back toward the noisy thoroughfare and opens to a courtyard that is shaded by the existing canopy of trees. The mass of the house shields and protects the courtyard from traffic noise beyond. The courtyard is the soul of the house with most of the rooms on the ground level opening directly onto it.

The act of discovery is deliberately orchestrated beginning with a broken pathway leading past a tall, sentinel-like stucco wall to the front gate of teak set into translucent glass sidelights. Beyond the gate one traverses a covered walkway enjoying views of the courtyard and the spaces that surround it en-route to the front door. Upon entering, a gallery is revealed that opens to a display of outdoor sculpture and thus punctuates the entry sequence. Finally from the main living space adjacent to the entry gallery an axial perspective of the courtyard is revealed and the entire concept becomes apparent.

By virtue of the large sliding doors from the living room and sliding pocket doors from the master suite and cabana, the indoor and outdoor spaces

become one. The flexibility of this arrangement makes the home perfectly suited to daily living as well as large scale entertaining by the owners.

This is the only modern home in this neighborhood and probably the largest. The dark, gray-green color is chosen in order to make the house blend into the existing canopy and to visually diminish its size. As completed, the house is a good neighbor, blending as planned with the trees and looking wonderful against the bright blue Texas sky.

本案附近是传统的别墅群，前面是高大而茂密的树林，后面是繁忙的街道。这些有利因素和局限因素正是设计这所住宅的出发点。

房子的平面呈“U”型，背后是繁华的街道，前面是被茂密的大树遮盖的庭园。宽大的住宅保护着庭院不受交通噪音的干扰。庭院是这所住宅的灵魂，一楼的很多房间都能直接通到庭院。设计精心，沿曲折的路径一路探索，经过高大的、哨兵一样的灰泥墙，可以通往镶嵌进半透明玻璃舷窗的柚木前门。穿过大门，可以在有屋顶的走廊上欣赏庭院的风景，周围的空间都与前门相通。进门之后，在走廊里可以看到室外的雕刻，点缀着入口。入口走廊相连的主生活区，与庭院的中心对应，整个设计概念得以清晰地显现。

客厅宽大的滑动门，主卧与浴室间的滑动折叠门，使得户内户外成为一体。布局灵活，适合主人日常居住以及举办大型娱乐活动。

这所住宅是周围唯一现代而且或许是最大的住宅。为了使住宅的色调与茂密的大树相协调，所以选择黑色、灰绿色为主打，以使它看起来纤巧。一旦完成，住宅会与周围融为一体，和树木相互衬托，在湛蓝的天空下，美妙绝伦。

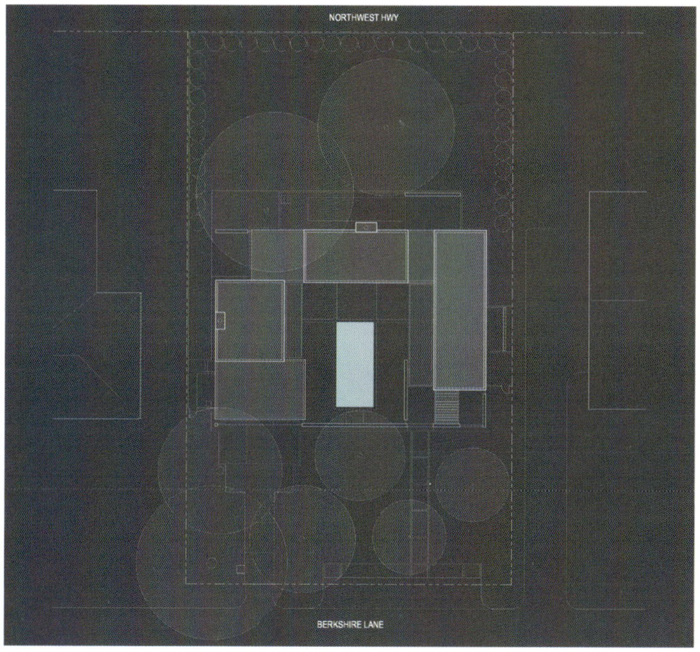

Tetris House

俄罗斯方块楼

Size: 550 m^2

总平面面积：550m^2

Designed by: Plasma Studio, Arch. Eva Castro, Arch. Holger Kehne, Arch. Ulla Hell

设计者：等离子工作室，建筑师伊娃·卡斯特罗，建筑师霍尔格·克内，建筑师乌拉·海尔

The building site is considered to sit on a potential archeological spot with remainigs from a roman settlement. Respecting the local limitation in hight development and wanting to realize the possible allowed size, the separted volumes are proposed. .

Multiple programmatic demands (five self contained units, parking space and exterior areas, potential archeological site) have initiated two volumes: L- shaped in plan and section they are spatially interwoven as to produce a range of in – between spaces.

The building is organized into two volumes which divide the "private villa" of the client with ist own indipendent access from the small unites which are rented out.

An open staircase between the two parts provides the access for the small units and the common spaces which are put in the underground level. Each volume cantilevers about five meters in order to provide covered parking. The cantiliver whose structural forces are legible in the wall and the ceiling planes is the result of a series of demands and restrictions from the local government on the one hand and the client on the other:

Local materials such as local larch wood are used throughout the design: in the outside as covering facade material, in the inside the wood is used with a white tint. The outside sloped and shaped façade continues in the inside as a built – in sideboard that turns into a series of steps and plinth beneath the internal stairs.

本方案多个项目要求（五个自给自足的单元、停车位和外部地区、潜在的考古地址）形成了两个建筑群：计划和截面中“L”形的应用，两个建筑群相互交织形成了一系列介乎两者之间的空间。

该住房分为两个建筑群，把客户有独立通道的 “私人别墅”与出租的小单元分隔开来。

两个建筑群之间的露天楼梯提供了通往地下一层的小单元和公共空间的道路。每个建筑群底下由大概5m的悬柱支撑，提供室内停车场。墙壁和天花板平面的悬梁柱的结构性力量清晰可见，一方面是当地政府一系列要求和限制的结果，另一方面是客户要求和限制的结果。

整个设计贯穿使用像当地落叶松这样的材料：在室外作为覆盖外墙的材料，室内用的木材染成了白色。室外倾斜成型的表面延续到室内，作为内置橱柜，变成了内部楼梯下一系列的阶梯和柱脚。

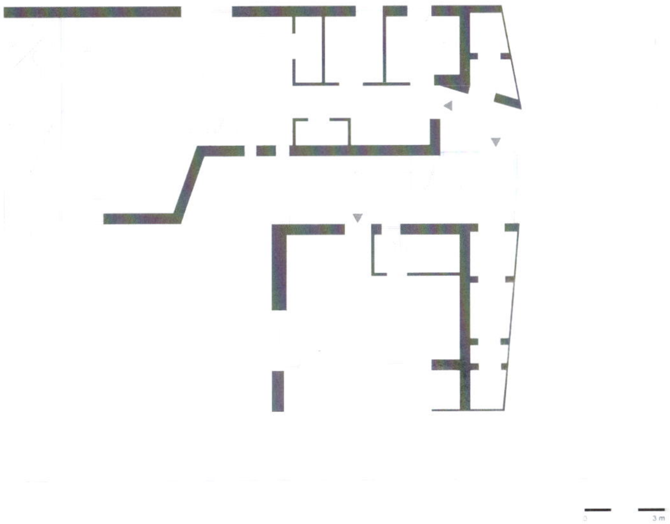

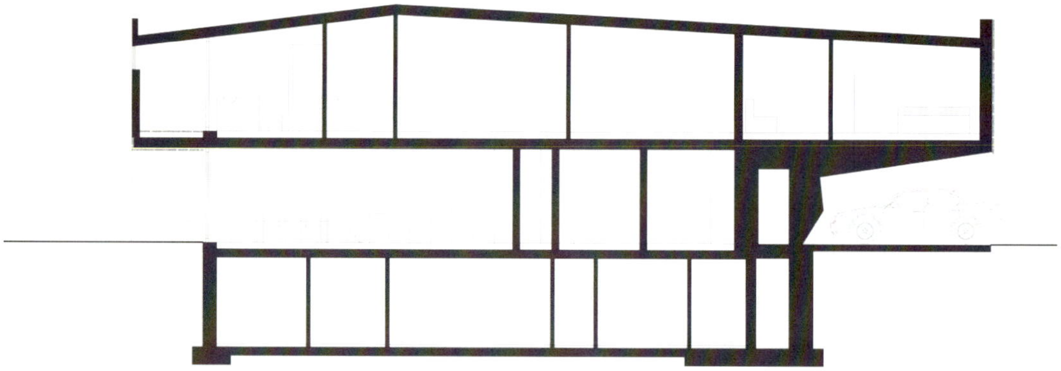
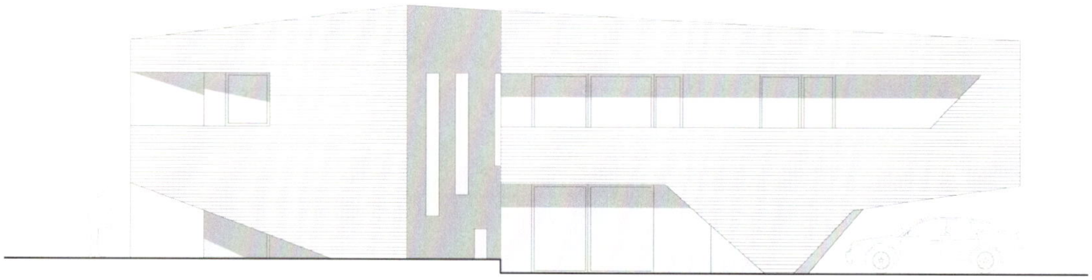

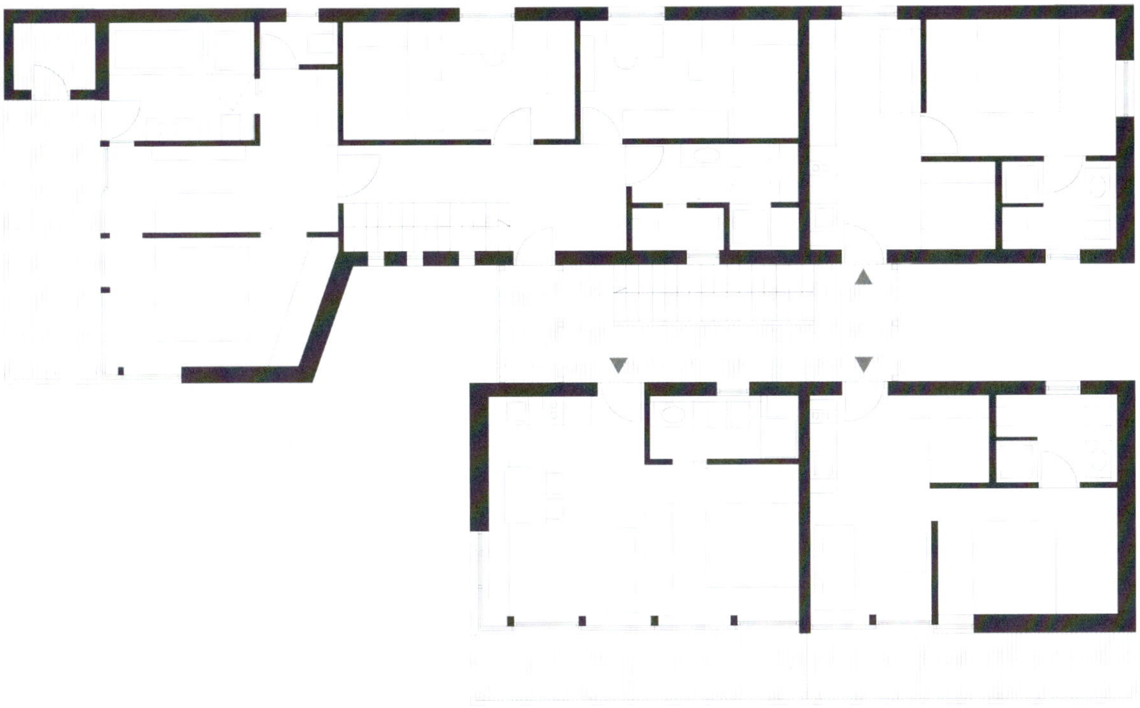

Single Family House

独户住宅

Architect: Nikos Koukourakis　Location: Dionysos, Athens, Greece

建筑师：尼科斯·库库拉奇斯　地点：希腊，雅典，狄俄尼索斯

The main architectural thought was to design a modern house with straight lines and big openings to a flat garden. This enables the natural light to travel through every space of the house, reducing the amount of energy needed to illuminate the house. Furthermore the house was designed and built according to "green" architectural choices of materials used, such as:

The continuous optical dialogue between the inside and the outside is achieved by the 3.50m (each panel) sliding windows that appear in the main elevation. The house was first designed in every detail in a computer 3-D model, which allowed to have a better understanding of the building's interaction with the environment.

The main interior design was made by the architect, as far as the fixed things of the house are concerned, such as staircases, fireplace, glass constructions, wooden floors, gypsum board ceilings and gypsum hidden lights constructions, lighting study, kitchen, bathroom tiles and cabinets.

The building dominates with its clean-cut lines, while its volume is split, without being dissolved, from the rich perimetric openings giving to it an airy sense via to their exposure to natural light. The big glass surfaces leave a permanently open dialogue with outside, without creating the least sense of isolation in the interior of the house.

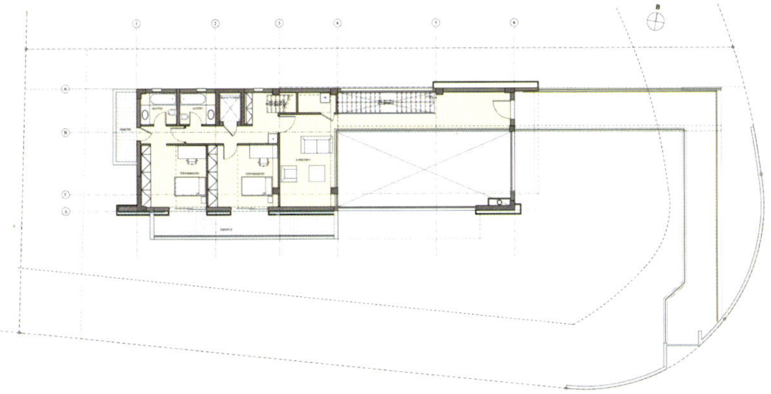

建筑的主要构思是设计一个直线造型的现代的住宅，有面花园一面完全敞开。这样自然光线可以进入住宅的每一个空间，节约了照明能源。此外，该房子的设计和施工选择“绿色”建筑材料，如：

在主要高度，3.50m（每个嵌板）的滑动窗口实现了内部和外部之间的连续光学对话。最初房子的每一细处是由电脑3D模型设计的，电脑3D模型能让人更好地了解建筑与环境相互作用的细节。

主要室内设计是由建筑师完成的，主要是房子的固定设施方面，如楼梯、壁炉、玻璃结构、木地板、石膏天花板和石膏隐藏灯建筑、书房照明、厨房、浴室瓷砖和柜子。

该建筑以简洁明快的线条为主导，四周丰富的开口带来大量自然光线，给人轻快的感觉，空间分隔开来，却也很宽敞。巨大的玻璃表面使房子与外界拥有永久自由的对话，房子内部没有任何分离的感觉。

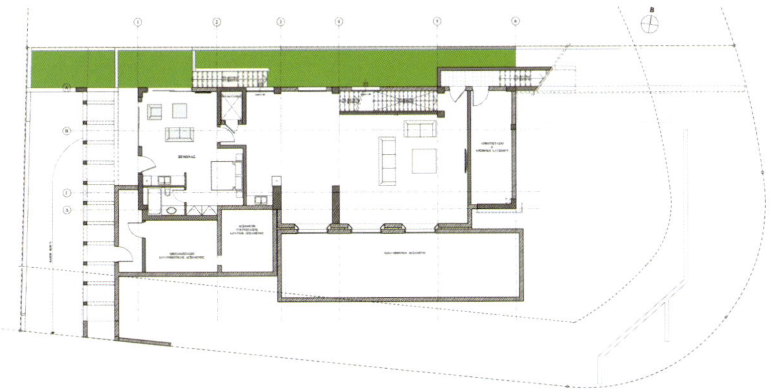

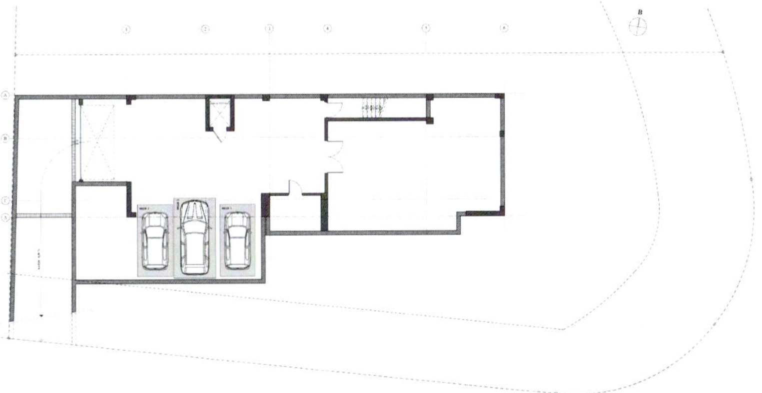

B House

住宅B

Surface building: 355 m^2 | Designed by: Damilano Studio Architects | Architects: Duilio Damilano, Claudia Allinio

地面建筑面积：355m^2 | 设计：达米拉诺建筑工作室 | 建筑师：Duilio Damilano, Claudia Allinio

The house extends horizontally with a row of premises that end up with the kitchen, real field glasses pointed to the village. The above mentioned volume dominates the access to the garage and controls the entrance to the property. In opposite directions, this house develops northwards with the body of the rooms more intimate and intimist and southward with the swimming pool, a wide sheet of water, reflections and light. The living floor is a spotless space, corrupt by the materiality of a fossil stone wall and by blobs of colour of some furniture. The first floor is reserved to landlords and hosts a wide room, the wardrobe box and the bathroom. The roofs are wide terraces, alternatives hanging by relaxation of the garden. In the night come out the depth of volumes and leaks highlighted by luminous cuts and clear/dark contrasts.

住宅横向延伸，房间成排，尽头是厨房，通过玻璃墙可以遥望村庄。上述空间占据着车库通道的两旁，控制着物业的入口。住宅向北是更加亲密和感性的房间，向南是游泳池。生活区，用化石石墙和带斑点的家具加以衬托。第一层有一个宽敞的房间是留给主人的，带衣柜和浴室。宽敞的屋顶露台，可以作为空中花园。当夜深人静的时候，明亮的房间和灯光与黑夜形成鲜明的对比。

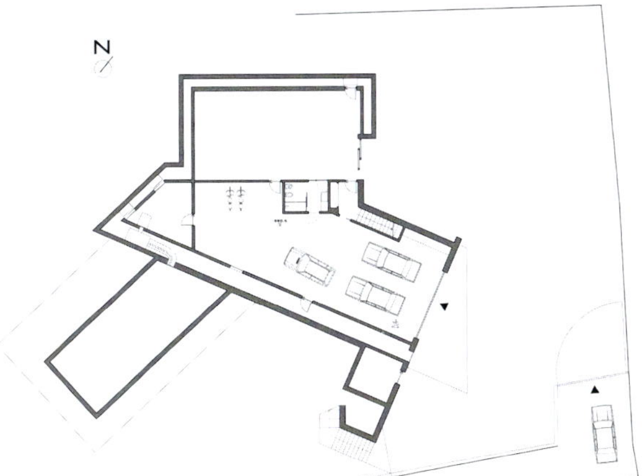

BASEMENT FLOOR

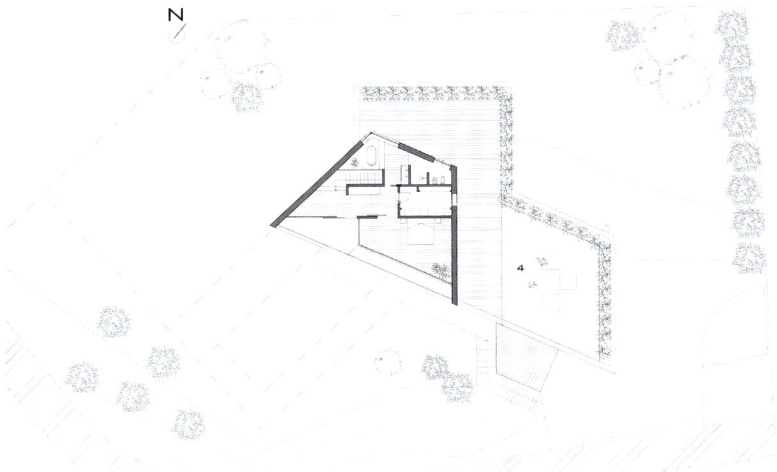

FIRST FLOOR

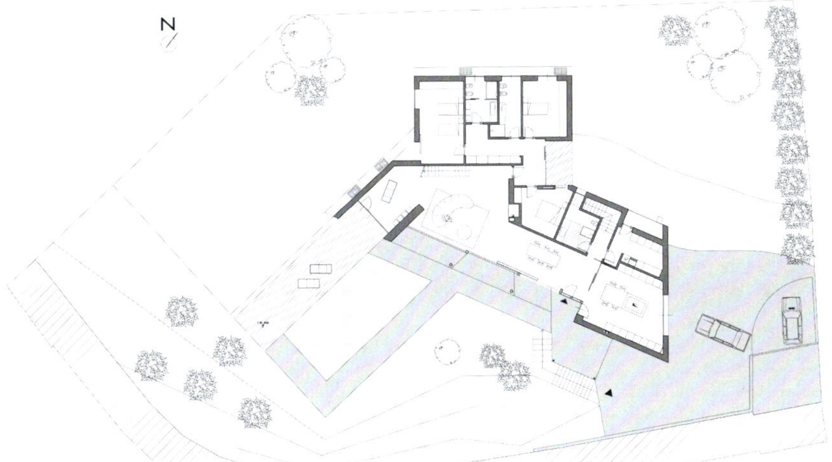

GROUND FLOOR

Horizontal Space

水平空间

Surface area: 2110 m² Designed by: Damilano Studio Architects

表面面积：2110m² 设计：达米拉诺建筑工作室

Horizontal Space is localized in Piemonte in northwest Italy a mountain area at 800 meters above sea level and the closest city is Cuneo.

Horizontal Space is a composition of pure volume of constant height, they moved horizontally and emptied or filled according to the function. A pomegranate highlights the access side.

Wide steps leading to the entrance introducing a longitudinal consecutive spaces.

Born to be divided into two units, the house is characterized by a large central hall, on the main breaks in the structure ca projected towards the outside with a wooden roof.

The living room, passing from one part of the house, opens with large windows to the garden and into the country. The teak floors warm environment and a stone wall covered with white sheets to create a staggered back almost sculptural material. The cuisine is characterized by a horizontal cut above the work plan, which frames the mountain and the nearby town of farms. In sleeping master a fifth stone fossil health by dividing the room and generates large shower with a satin flat floor. A translucent sliding island ideally the bathroom.

The villa also has two bedrooms with their bathroom, a laundry and a bathroom for the living area.

The front of the rear, much more austere, has a balcony of the service and the opening of the premises in the basement used as a gym, garage and local technicians.

水平空间坐落于意大利西北部的皮埃蒙特，位于一个海拔800米m山区，最靠近的城市是库内奥。

水平空间布局纯粹，高度恒定，房间水平分布，可以根据用途需要进行改建。一棵石榴树点缀着入口一侧。

沿宽阔的台阶而入，便是一个狭长的，流畅的空间。

住宅自然地一分为二，宽大的中央大厅是这座住宅的特色， 向外凸出的木顶，是这座住宅结构的主要突破。

住宅的一角是客厅，宽敞明亮的窗户通向花园，融入美丽的乡村风光。柚木地板给人以温暖的感觉，石墙饰有白色木板，参差交错，宛若雕刻。厨房的特色是在烹饪台上方有一个横向缺口，可以欣赏山脉和附近城镇农场的风景。将主卧隔出1/5的地方，用石头铺成一片保守的保健区，装上大型淋浴喷头，铺上光滑的地板。半透明的滑动门完美地将卫生间隔开。

这座住宅还有两个带卫生间的卧室，生活区有一个洗衣房和一个卫生间。
后部的正面，更加简洁，有一个可以招待客人的阳台，以及作为健身室、车库以及存放专业工具的房间的出口。

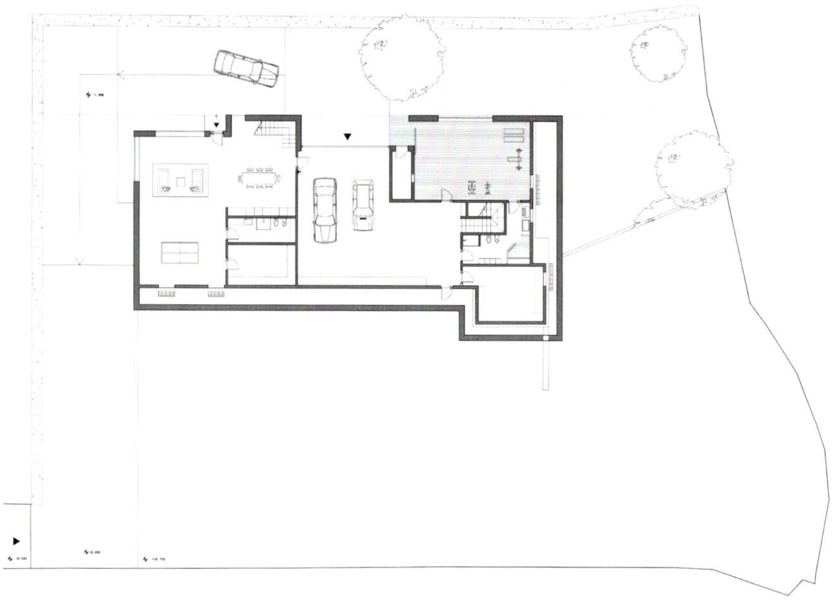

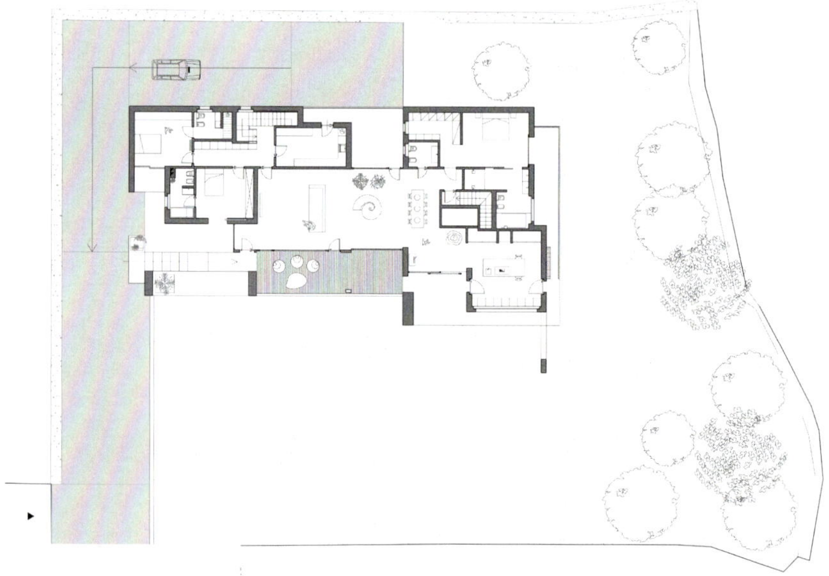

House KVD

住宅KVD

Gross floor area: 432 m²　Architect: Grosfeld van der Velde Architecten

总建筑面积：432m²　建筑：Grosfeld van der Velde Architecten

House KVD is located in the woods of Teteringen, close to Breda. Important, from the residents point of view, was to experience the green environment in a most intense way. This ambition has formed the main concept of the building: A house that feeds in between the trees, giving every arbitrary movement a surprising perspective on the forest. The square-shaped volume keeps you, while moving through the building, in constant contact with the environment.

The square, that's partly lifted by piles, embraces an open courtyard on which the entrance is situated. The living rooms are located on ground zero, the bedrooms are lifted and provided with horizontal orientated windows. The framework, situated on the outside of the building, are filtering direct incoming sunlight and give a beautiful 'cut out' view of the forest. They also intensify the sculptural characteristic of the volume. The façade is made of vertically placed black stained wooden parts. The subtle details in combination with shape and material make the house modern, but also reserved in it's afforested environment.

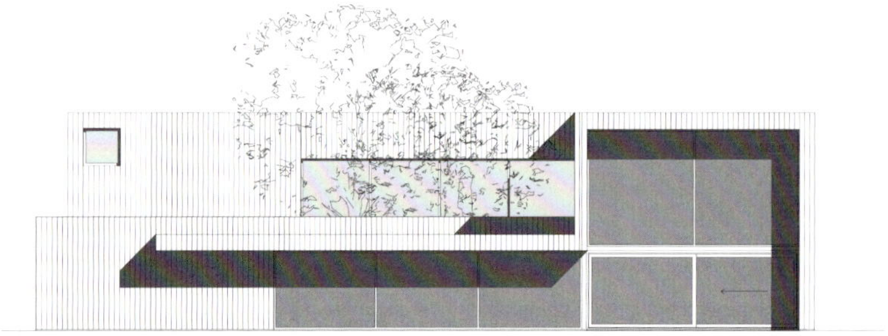

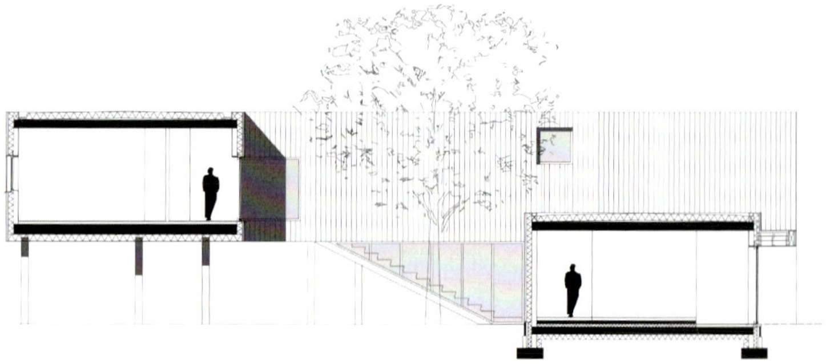

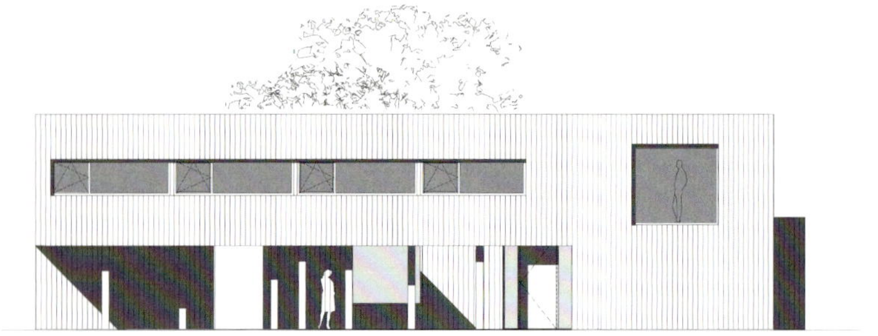

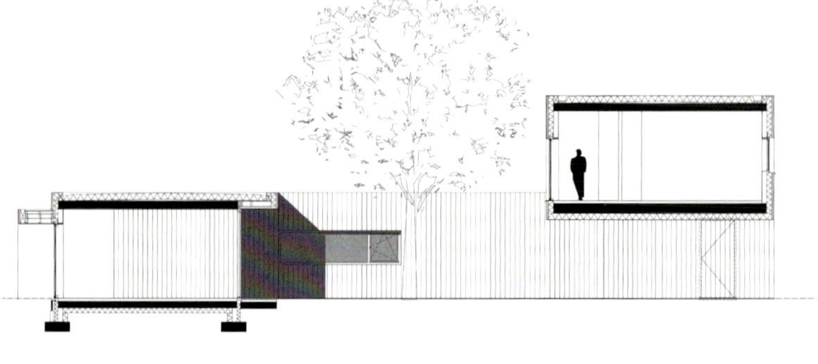

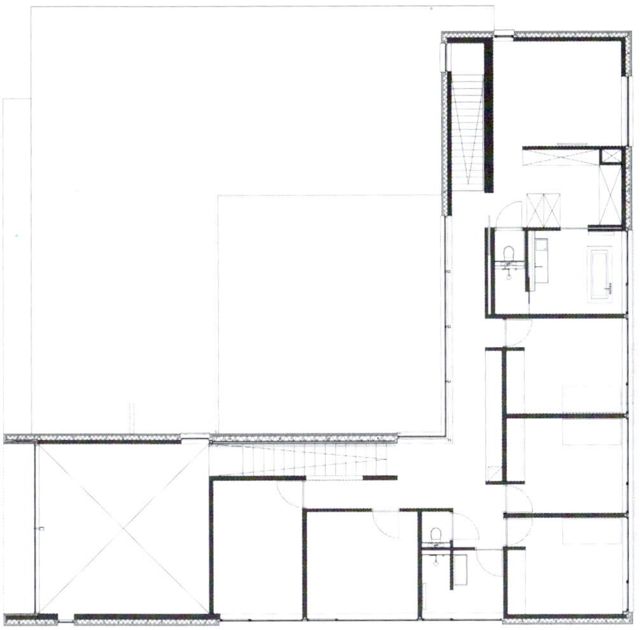

10. gangzone
11. werkruimte
12. werkruimte
13. toilet
14. badkamer
15. slaapkamer
16. slaapkamer
17. slaapkamer
18. badkamer
19. toilet
20. walk-in closet
21. slaapkamer

10. corridor zone
11. office
12. office
13. toilet
14. bathroom
15. bedroom
16. bedroom
17. bedroom
18. master bathroom
19. toilet
20. walk-in-closet
21. master bedroom

0 5 10

住宅KVD坐落在泰特灵恩森林里，靠近布雷达市。重要的是，居住者可以亲身体验并尽情享受苍翠的森林美景。这一奢望也是该建筑的主要理念：一所住宅掩映于树林中，任意游走，都能纵览森林令人惊讶的景观。方正的造型让你无论处在房屋的哪个角落，都能与大自然融为一体。

这一方方正正的建筑，局部用桩柱支撑，怀抱一个开放的庭院，入口就位于这里。

客厅在一楼，卧室在高处，有横向的窗户。建筑外部的架构，透过直接照射进来的阳光，映射出森林美丽的“剪影”。它们也使这一空间雕刻般的感觉得以强化。外墙采用竖的黑色木板。形态和材质的微妙结合使得这所住宅在浓郁的森林中时髦而又含蓄。

HOUSE S

住宅S

Gross floor area: 518 m^2 Architecture designed by: Grosfeld van der Velde Architects

总建筑面积：518m^2 建筑设计：Grosfeld van der Velde Architects

House S is located on the edge of the Asterdplas in Breda's Haagse Beemden district. The living area is raised above the surrounding land, so that the view on the natural environment is optimal. Further solidifying the home's connection to nature, large floor-to-ceiling windows frame the view, while giving the house a light and transparent character.

It aims to design a house that reacts primarily on the spectacular views from the plot. The house is embedded in a typical but boring Dutch urban planning environment and we wanted to make a statement. There were a few plots on which we could transform the site into a totally different environment. The house and its garden give the impression of a film set. The challenge in this house is to incorporate the minimalism of the architecture in the beautiful and green environment.

Since the residents are home workers, employees and customers are also regaled at the house. Therefore private and business rooms are separated in an ingenious way, where they do touch but never cross. The working space is partly situated below grade, fit with a horizontal orientated window. Providing a beautiful view on the environment and allowing for a bright, natural lit lower level.

The house is made of stone and concrete bearing walls. The façade is made of smooth plaster, and is painted grey-white. The terraces are made of hardwood elements, the window frames are aluminium and the cladding of the garage and the entry is made of western red cedar wood. The flooring is all made of white epoxy.

住宅S坐落于布雷达市Haagse Beemden区Asterdplas的边缘地带。生活区高于周遭陆地，可以欣赏最优美的自然风景。住宅更加强化了住宅与自然的融合，大型落地窗将风景纳入眼底，而且使得住宅明亮而通透。

设计这座住宅，旨在和这里的壮丽风景相应成辉。我们想声明一下，这座住宅地处一片典型而繁琐的荷兰城市规划区域中。有几块地块，我们可以在此基础上将本案改变一下，变成完全不同的风格。这座住宅和它的花园看起来像是摄影场地。设计这座住宅所面临的难题就是如何把建筑简派艺术融入美丽而苍翠的风景中。

因为居住者在家工作，所以雇员和客户也经常往来于这所住宅。因此将私人房间和工作房间巧妙地隔开，它们相互衔接，但是互不影响。工作区有一部分是在楼下，装有横向的窗户。透过窗户，可以欣赏美丽的风景，也能让楼下接受明亮的自然光的照。

住宅用石材建造，水泥承重墙。外墙的材料是细腻的灰泥，油漆成灰白色。硬木露台，铝合金窗架，车库的外层和入口都是西方红雪松材质。地板全部采用白色的环氧树脂材料。

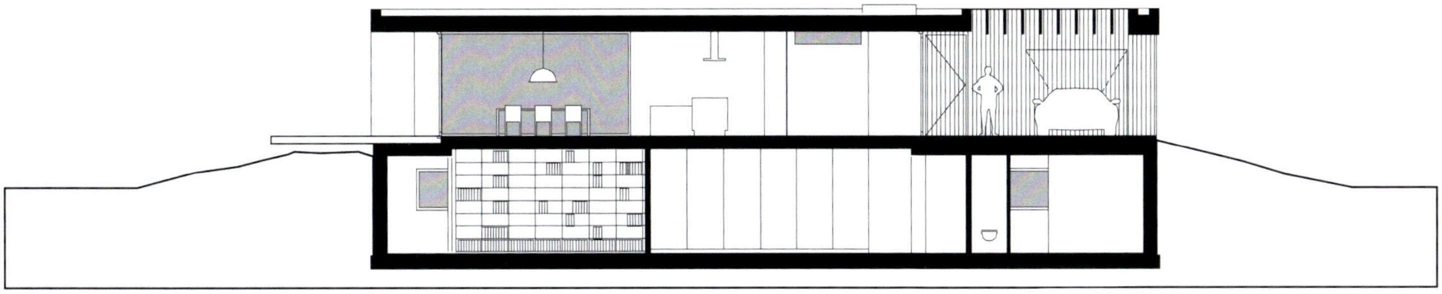

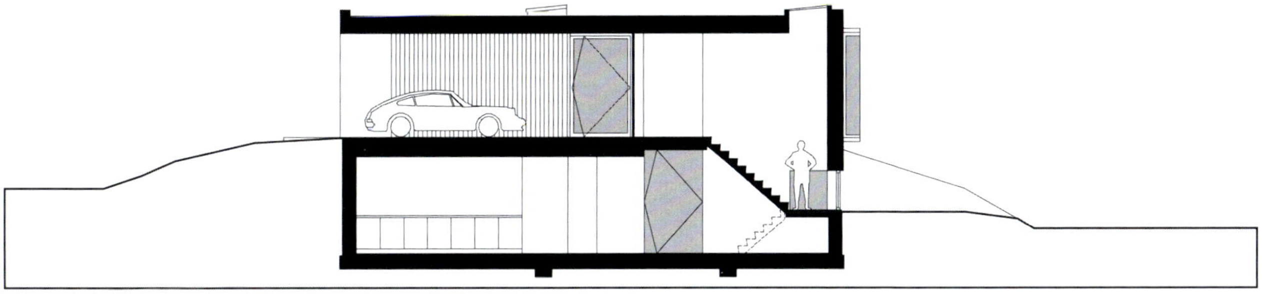

1. living room
2. dining room
3. kitchen
4. storage
5. master bedroom
6. cloakroom
7. private entrance
8. business entrance
9. garage
10. carport

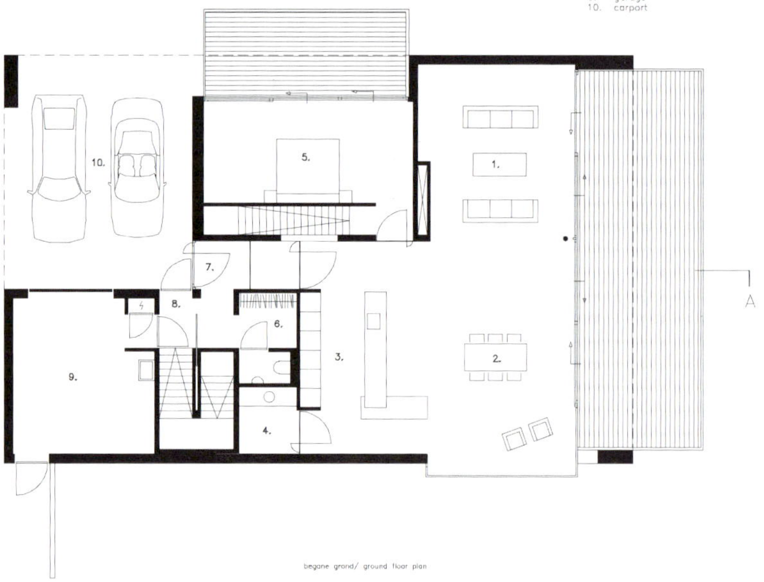

begane grond/ ground floor plan

11. laundry room
12. master bathroom
13. walk-in closet
14. guest bathroom
15. guest bedroom
16. pantry
17. office
18. storage
19. meeting room

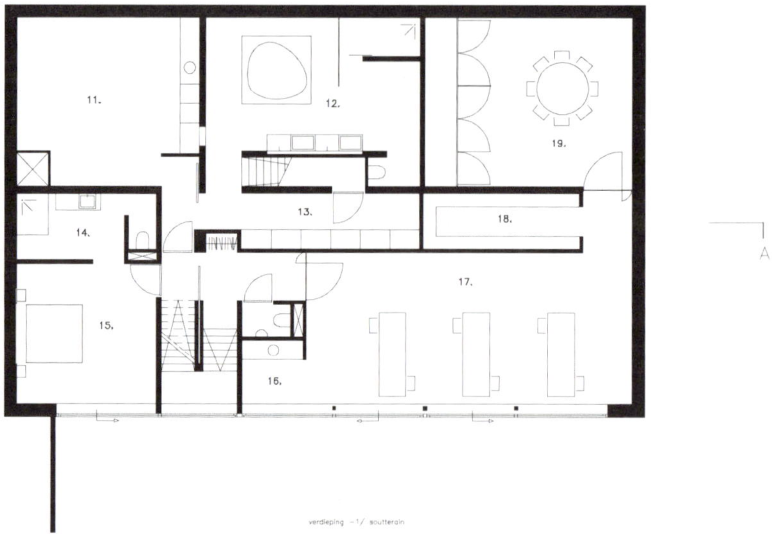

verdieping -1/ souterrain

House like a Museum

博物馆式住宅

Site Area: 776.4 m² Architects: Edward Suzuki Associates (Edward Suzuki, Toshiharu Nanba)

总平面面积：776.4m² 建筑：爱德华·铃木联合公司（爱德华·铃木，Toshiharu Nanba）

Located in only about a five-minute walking distance from the railroad station, this house sits in a commercial/residential zone. As such, the neighboring houses and shops encroach upon the site rather closely, so much so that from the start the idea to "look out" from the property was abandoned to instead "look inward" was adopted. Since the client-owner had a vast amount of art collection and wished to display such art in the new house, the theme "House like a Museum" became the design generator.

Near the entrance and up on a mezzanine in a semi-detached fashion is a Japanese Room able to accommodate occasional guests. Next to it and readily accessible from the entrance is the owner's Private Office. Leading up from the entrance is a sloping gallery-corridor connecting to the Living, Fireplace, Dining, and Kitchen areas. Again, another ramp connects this one-room like space to the elevated Master Bedroom with its private water closet. From this elevated area a staircase leads to the Family/Home Theater space adjacent to the Children's Bedrooms and Family Bathroom on the second level. On the south of these bedrooms spread a spacious garden terrace overlooking the circular patio below.

这座住宅地处商业住宅区，离火车站只有5分钟的步行路程。所以，邻近的住宅和商店都非常接近，甚至从这所房子你刚想"往外看"就已经放弃了这个念头，就又"朝里看"了。因为业主有很多艺术收藏品并且希望能在新居把这些艺术展示出来，所以"博物馆式住宅"这一主题就是设计的出发点。

入口附近，在夹楼中有一间半独立的日式房间，可以供临时客人住宿。它的隔壁是主人的私人办公室，从入口即可方便地出入。从入口处向上，是一个倾斜的走廊，连接着客厅、壁炉和餐厅以及厨房。此外，还有一个斜梯，连接着太空间，给抬高的带私人卫生间的主卧预留了一定空间。从这一抬高的区域，有一个楼梯通往二楼的家庭中心空间，它隔壁是儿童房和家庭浴池。在这些卧室的南边，是一个宽大的花园露台，俯瞰着下面圆形的天井。

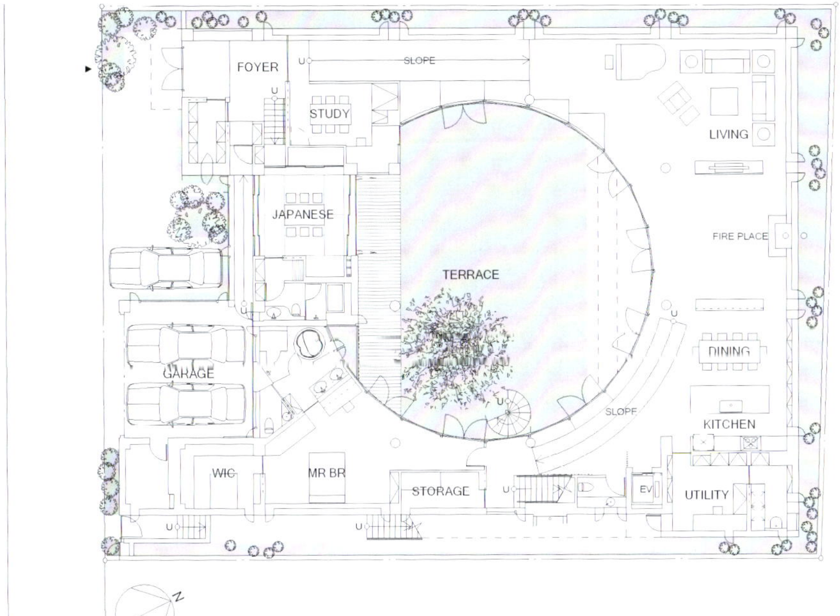

1 F　419.85㎡

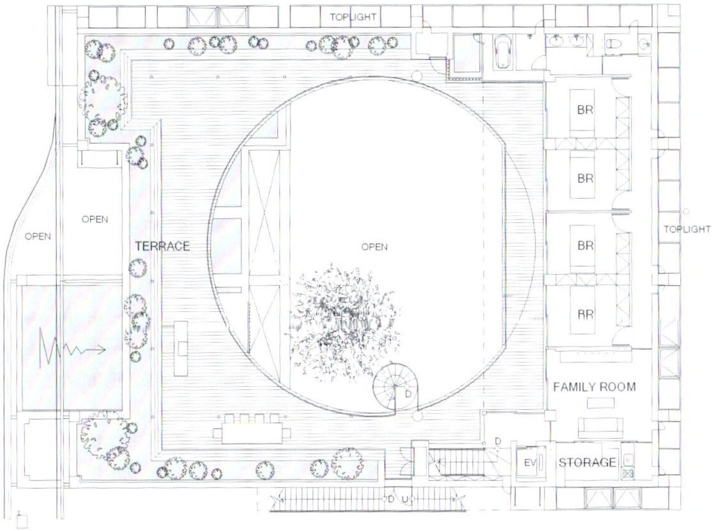

2 F　118.49㎡

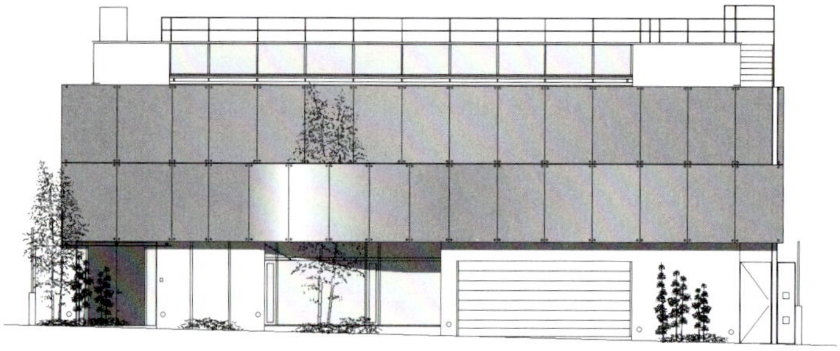

ELEVATION

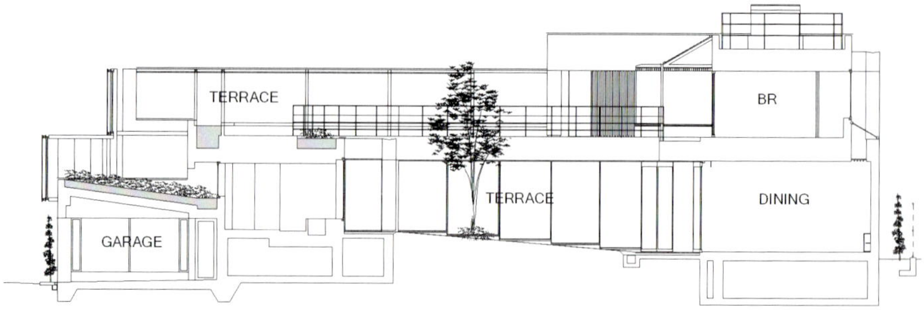

SECTION

Ehrlich Residence

Project Team: John Friedman (principal-in-charge), Alice Kimm, Leigh Christy, John Martin, Pauline Shu

埃里克居

建筑：约翰・弗瑞德曼・爱丽思・金姆建筑师事务所(John Friedman Alice Kimm Architects)

The resulting first floor is an L-shaped series of living spaces wrapped around the garage and opening onto an L-shaped garden, though the corners of both L's are dematerialized. On the southeast side of the house, eaves block the summer sun but allow the winter sun to heat up the concrete floor, which contains radiant heating and acts as a heat sink. The southwest façade is mostly solid to keep out the punishing western sun. With no air conditioning, the reflecting pool cools the breezes that enter the living room. The house is clad with two sustainable materials: plaster and cement board. The garage door is clad with stainless steel, while the front door, like the wood on the interior stairs and second floor, was sustainably harvested.

The atrium has other important roles. First, its verticality contrasts with the horizontality of the first floor – a horizontality that intensifies the interior's relationship with the garden, which is planted with drought resistant plants and partially irrigated by gray water. Secondly, its mixture of reflected and direct light effectively tracks the sun during the daytime, and at night emits a soft glow. Finally, the atrium organizes the rooms on the second floor, particularly the master bedroom, which has an internal window that allows the clients to take in the ever-changing light in the double-height space, as well as see someone below. In addition, it was the pinching of the true south and north corners of this floor that produced the location for the monitor and the kitchen skylight, and also made it possible for a fat, square plan to have more of the qualities of a thin plan. Situated to block noise from the busy boulevard, the master bathroom takes advantage of this quality with light from three different orientations.

该住宅环绕着车库，一楼是“L”型的生活空间，朝向一个“L”型的花园，但两个“L”型的转角都不太明显。住宅的东南侧，屋檐阻挡了夏日的烈阳，但在冬天，太阳能照射在水泥地板上，地板里装有辐射供暖系统，担当散热片。西南外墙是坚实的墙壁，以遮挡西晒。无须使用空调，因为掠过倒影池，凉爽的微风可以吹进客厅。这座住宅的外层使用两种可持续材料：石膏和水泥板。车库门是不锈钢材质，而前门，与内部楼梯和二楼的木材一样，是可持续材料。

中庭还有其他重要的作用。首先，它的垂直和一楼的水平状态形成鲜明对照——水平状态强调了内部与花园的关联，花园里种植的是耐旱植物，有些是用中水进行浇灌的。其次，它既吸收反射光又吸收直射光，白天能有效地接受太阳的光照，到晚上就散发出柔和的光辉。最后，中庭使得二楼的房间显得井井有条，尤其是主卧，它有一个内窗，能让主人注意到二楼灯光的变化，也能看到下面的人。此外，这一楼层正南正北转角收聚，给监视器和厨房天窗预留了位置，并且使得方正圆润的布局显得更加简洁可爱。作为屏蔽繁忙街道喧嚣的要塞，主卧卫生间三面都装有照明设备。

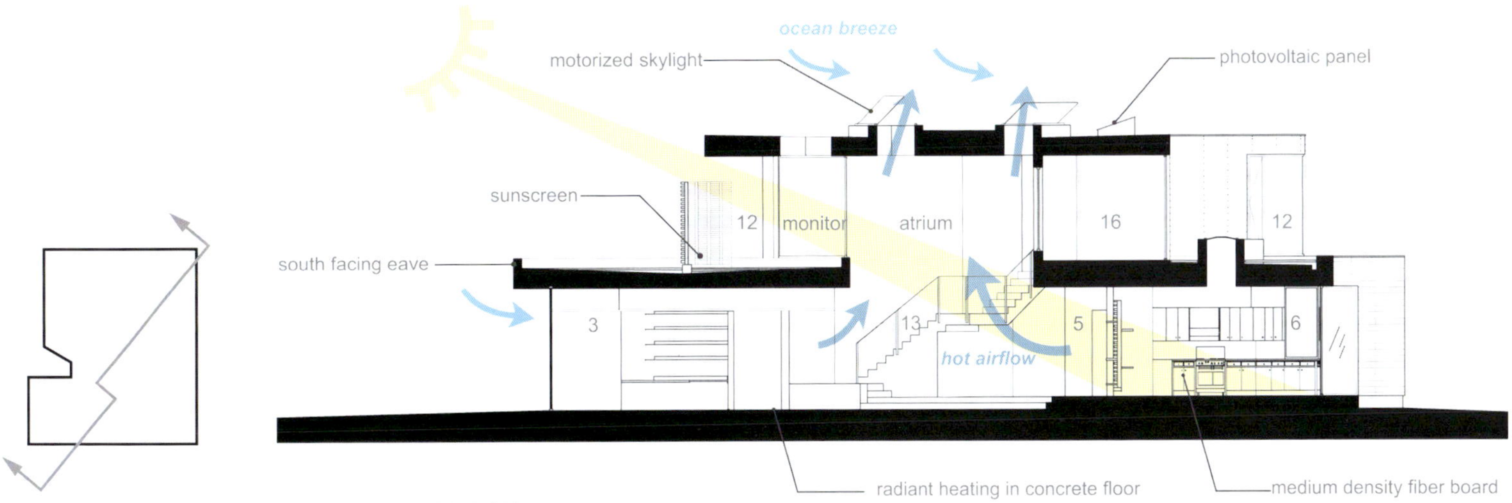

DIAGONAL SECTION
void / atrium

3 office
5 dining
6 kitchen
12 terrace
13 stair / atrium
16 master bedroom

SITE AND FIRST FLOOR PLAN

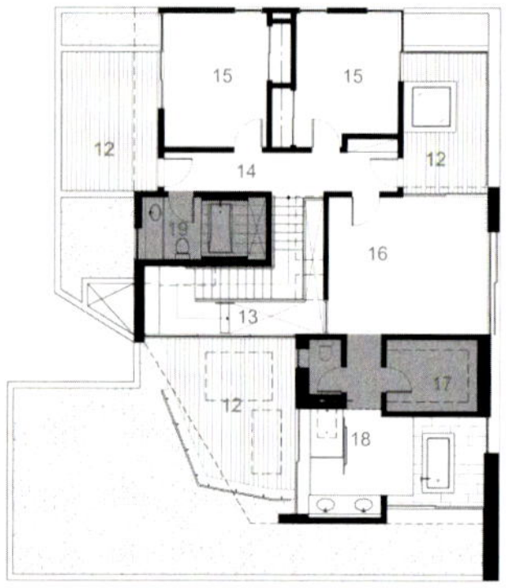

SECOND FLOOR PLAN

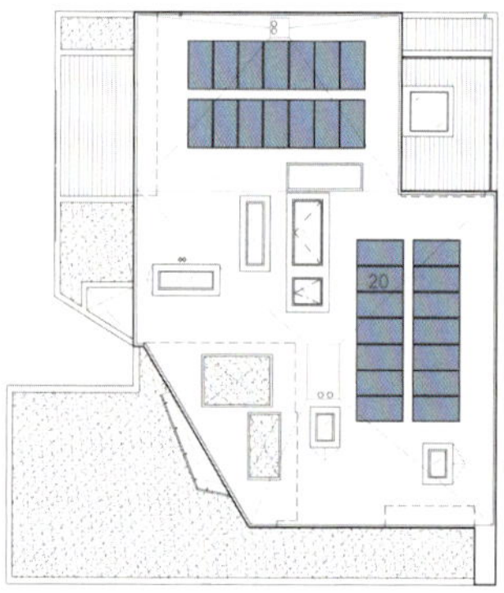

ROOF PLAN

House J2

住宅J2

Site area: 687 m²　　Designed by: 3LHD architects

用地面积：687m²　　设计：3LHD建筑事务所

The family house is located in the green residential part of the city of Zagreb. On both sides the site is bordered by a street and a high building. These contextual facts determined the concept and the shape of the new project. The "L" layout with closed fronts, protects the house from the street and the neighbouring building. At the same time the garden has been redesigned with all the main rooms in the house oriented towards it.

The living room, dining room and kitchen form a unique space and together with a swimming pool are built into the ground. In this way, being at the same level and separated from each other by a glass wall they bring the garden into the house. The house entrance is above, at street level, together with garage, storages, closet-space and studio. The family area is above the entrance space along with the living and dining rooms.

The materials used for the façades correspond to the spatial organisation of facilities. The living and dining spaces are separated by glass walls which completely open the living space to the outside; on the other hand, the bedroom walls are alternatively panelled by wooden boards.

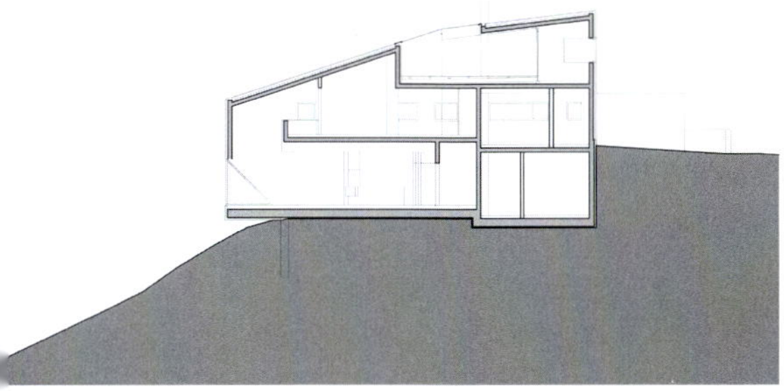
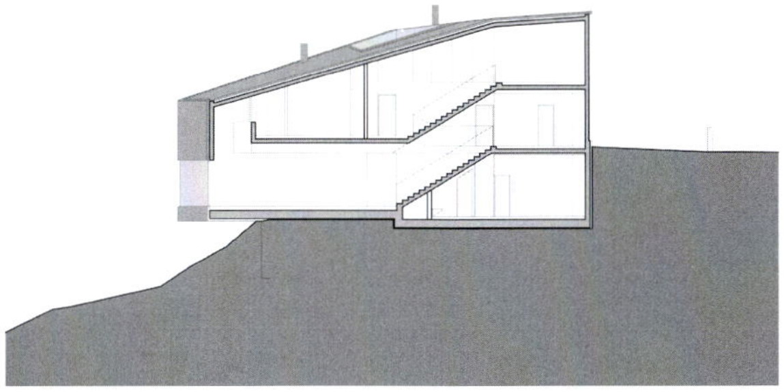
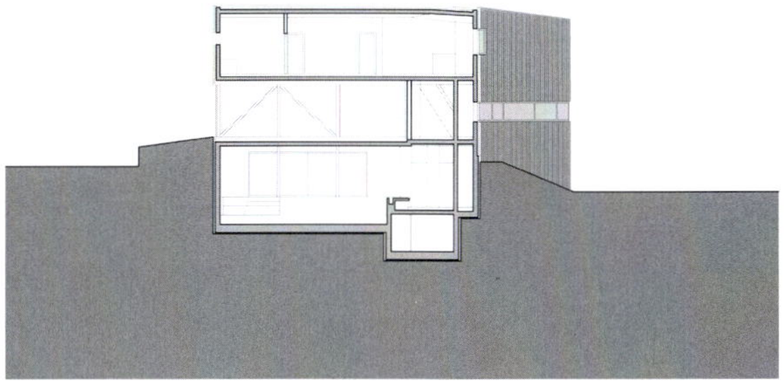
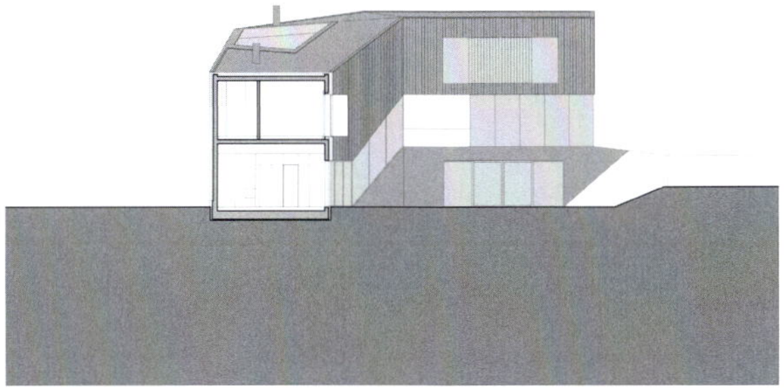
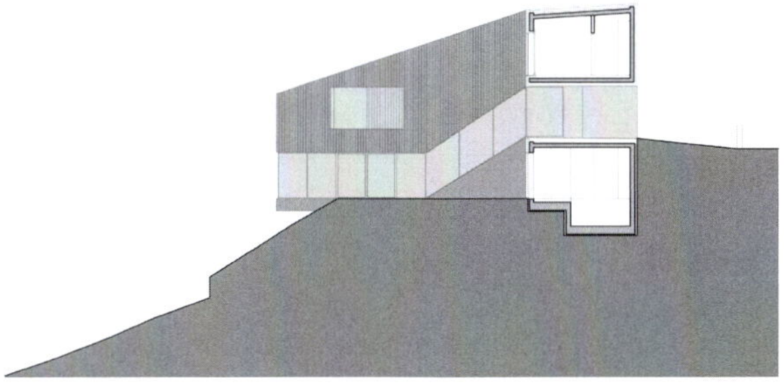

1 Covered parking
2 Garage
3 Storage
4 Entrance
5 Wardrobe
6 Wc / shower
7 Guest room
8 Master bedroom
9 Bathroom
10 Gallery
11 Staircase

12 Kitchen
13 Dinig room
14 Living room
15 Storage
16 Laundry /boiler room
17 Sauna
18 Trim
20 Pool

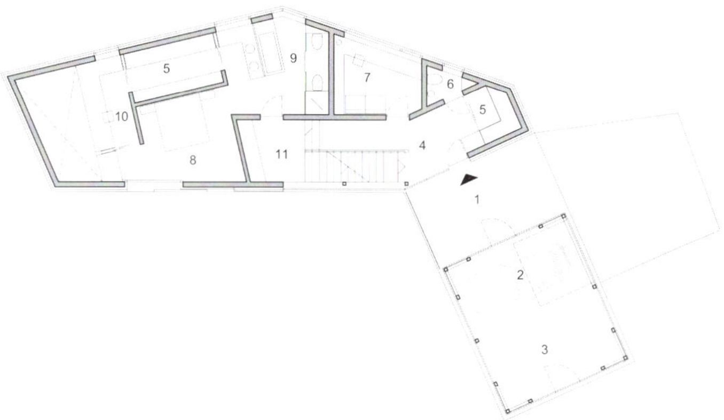

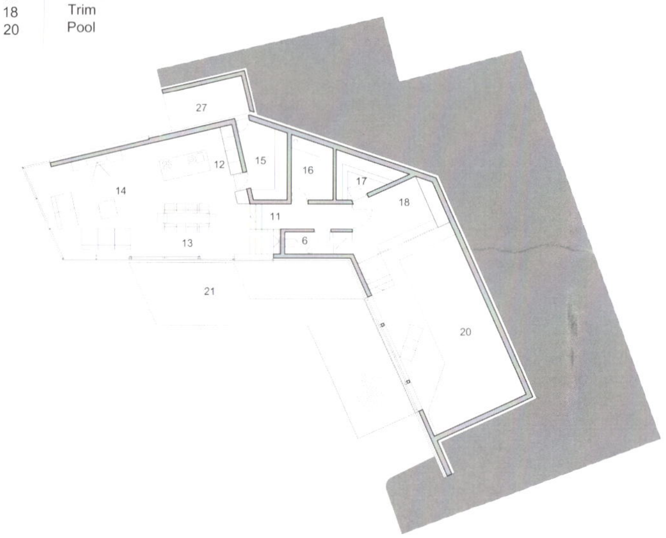

23 Living- children
24 Closet
25 Room
26 Bathroom

21 Terrace
22 Garden
27 Utility terrace
28 Barbecue

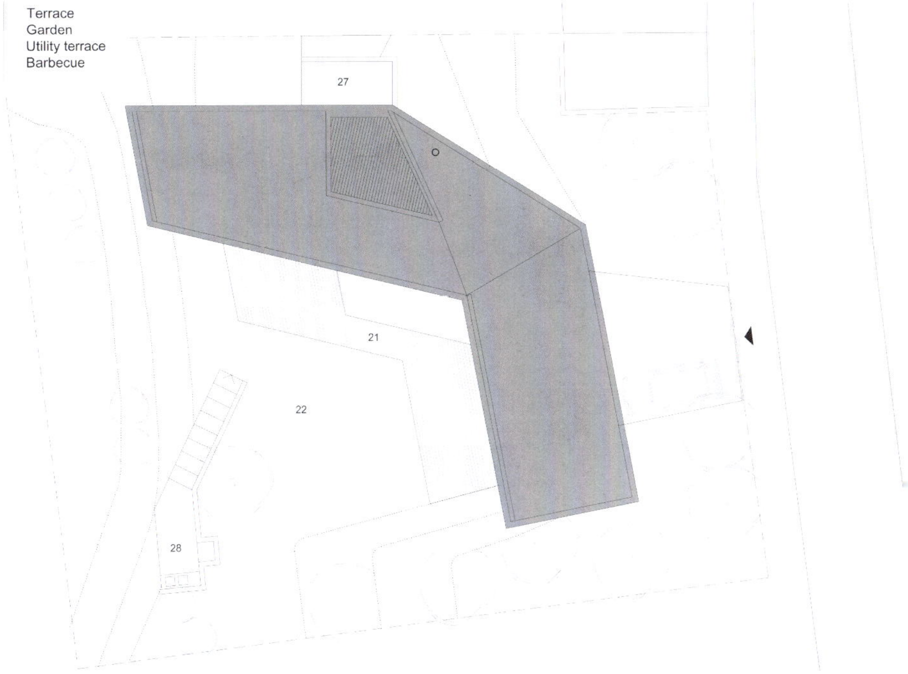

本住宅坐落于萨格勒布市青翠的住宅区。本案两旁临街，高层建筑林立。这些背景情况决定着新项目的理念和造型。“L”型布局，前墙封闭，避免住宅受到街道的喧闹和邻近建筑的影响。同时，花园经过重新设计，使得所有主要的房间都能通向花园。

客厅、餐厅和厨房，构成了一个独特的空间，连同游泳池，都在一楼。这样，在同一水平面上，一扇玻璃墙将它们彼此隔开，花园风景，便尽收眼底。住宅入口高，与街面相平，带有车库、储藏室、密室和工作室。家庭活动区高于住宅入口处，与客厅和餐厅相连。

外墙用材符合设施的空间组织原理。客厅和餐厅用玻璃墙隔断，将生活空间完全敞开；另一方面，卧室墙壁则采用木质墙板。

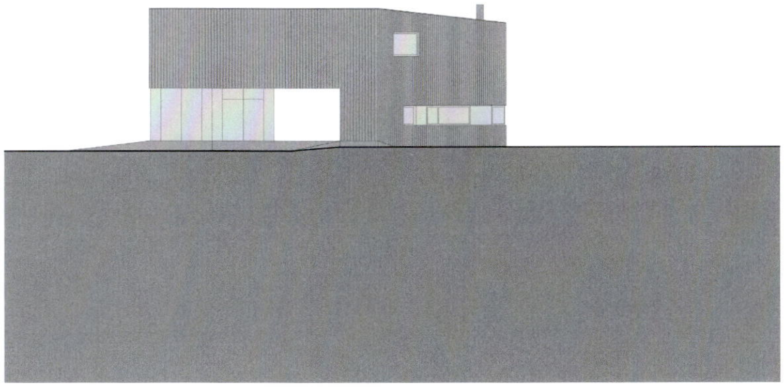
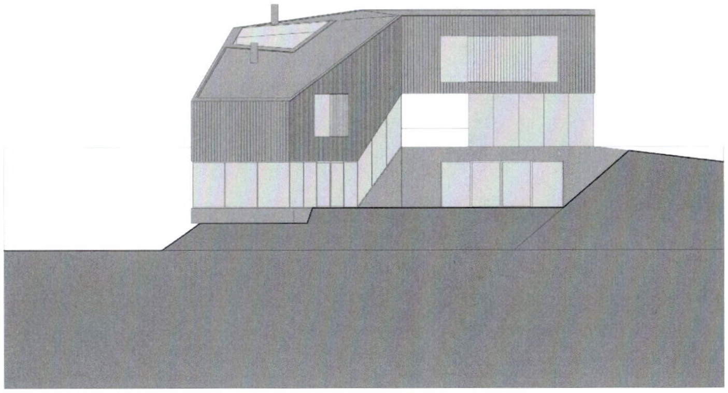
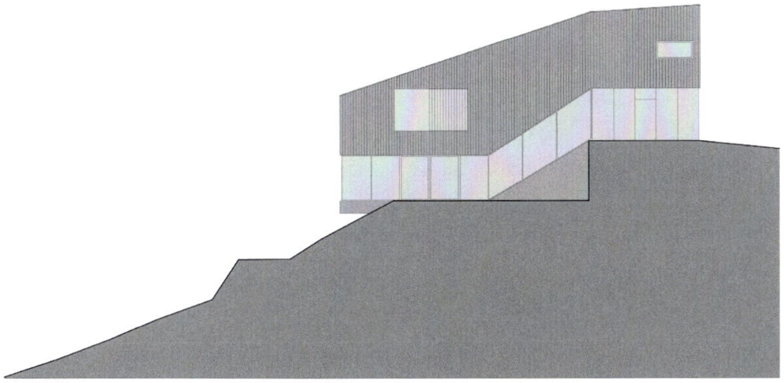
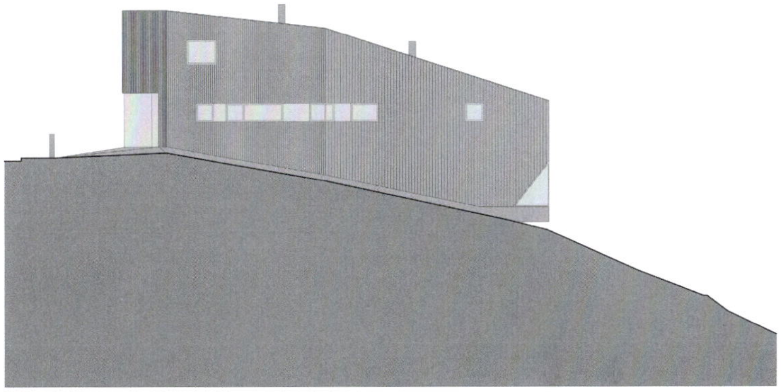

House K

K家

Site area: 1110 m^2 Designed by: 3LHD Architects

用地面积：1110 m^2 设计：3LHD建筑事务所

Family house K is situated on a large residential plot on a moderate eastern-facing slope at the edge of a forest near the base of Sljeme hill. By placing the object next to the street in the north-western corner of the plot, preference was given to a beautiful forest view; the uneven terrain was used as a means of developing the object's volume.

The concept of the house is inspired by the shape of a snail's shell, which involves a spiral twisting around its own axis, and as such all functional elements of the house follow each other and are built around a central atrium. The atrium was made in order to compensate for a lack of sunlight in the living-room area and enabled two-way lighting of certain rooms.

The house is maximally open and rises two floors towards a vast green meadow and forest; towards the street the house is enclosed by a simple ground-floor facade with a hidden garage door.

Almost in opposition to the progression of a snail's home, the modest and unnoticeable entrance K House is situated at the edge of the southern façade. This is where the spiral movement around the atrium and through the half-levels begins; it ends in a big living room that opens towards the outdoor terrace and the forest.

Diverse and colorful elements which find a place in the home in order

1 Living room
2 Hobby
3 Kitchen
4 Wine storage
5 WC
6 Storage
7 Sauna
8 Trim
9 Swimming pool
10 Atrium
11 Workshop
12 Garage
13 Pool Terrace
14 Living room terrace

15 Entrance
16 Bedroom
17 Bathroom
18 Wardrobe
19 Boiler room
20 Laundry
21 Living - children

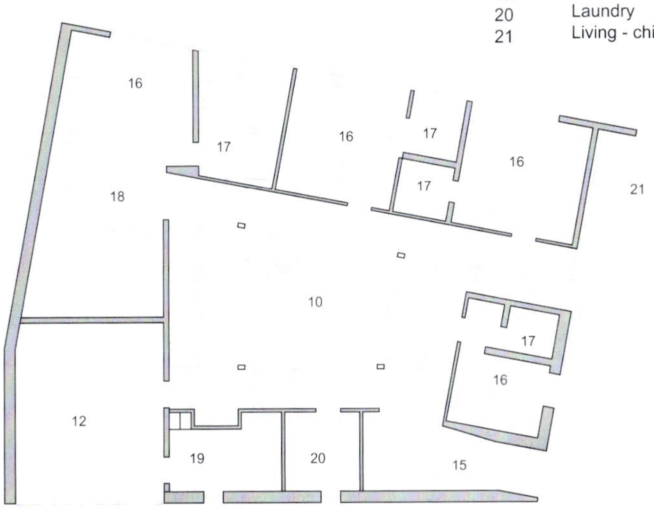

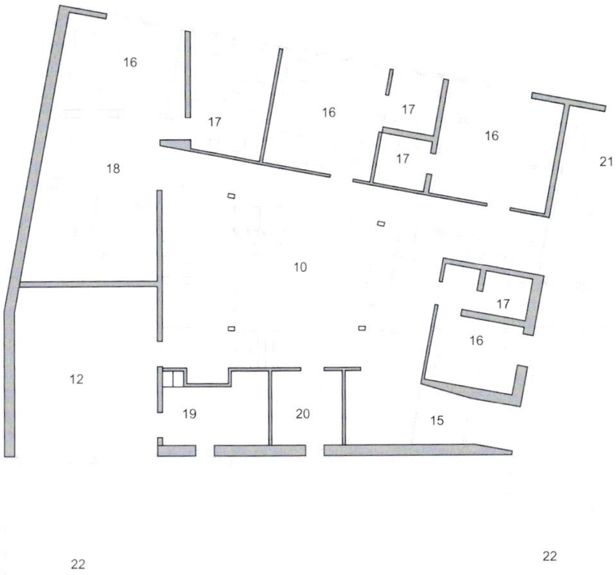

15 Entrance
16 Bedroom
17 Bathroom
18 Wardrobe
19 Boiler room
20 Laundry
21 Living - children
22 Entrance path

to satisfy the needs and wants of the inhabitants are developed in a 600 m2 area and in several half-levels, thus making the best possible use of the terrain's difference in altitude and visually making the house appear smaller.

All facades and terraces are paneled with untreated teak boards. The use of wood paneling continues in the interior in response to the spiral progression, as well as to achieve an intimate and comfortable atmosphere. Therefore floors and, intermittently, walls are paneled with teak wood while ceilings and walls are primarily white. Black appears as an additional color accent on the walls, offering a means of directing one's gaze and movement.

An interesting alternative to the straightforward coloring of the home interior was brought to bear in the wellness and swimming pool area. In this dim ambience with a predominantly black lattice ceiling and a wavy golden wall, true atmosphere is created by changeable RGB lighting ranging from purple to blue and giving the area an almost set design-like dimension.

A curious addition and final accent to the house were made by the roof terrace. One reaches it via an external spiral staircase that hangs in the atrium between the first level and the roof, thus accentuating the fact that a family house, besides being a home for a family, has also to serve as a place for indulging in companionship, comfort, luxury and gratification.

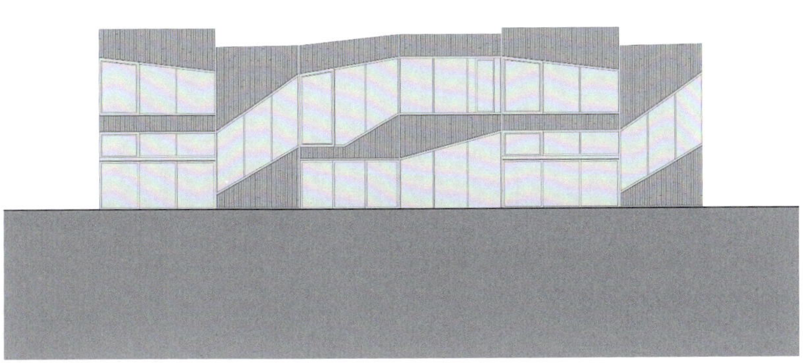

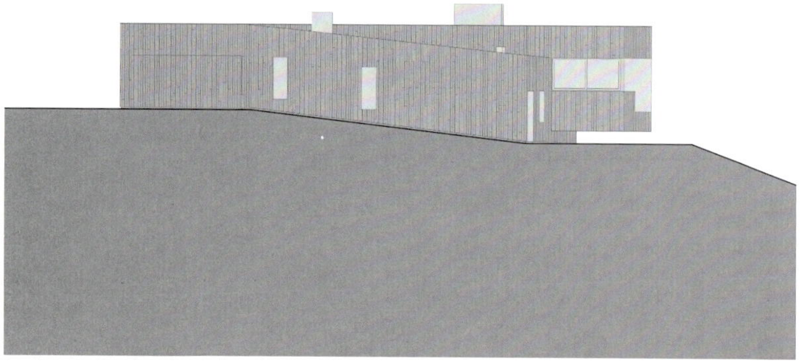

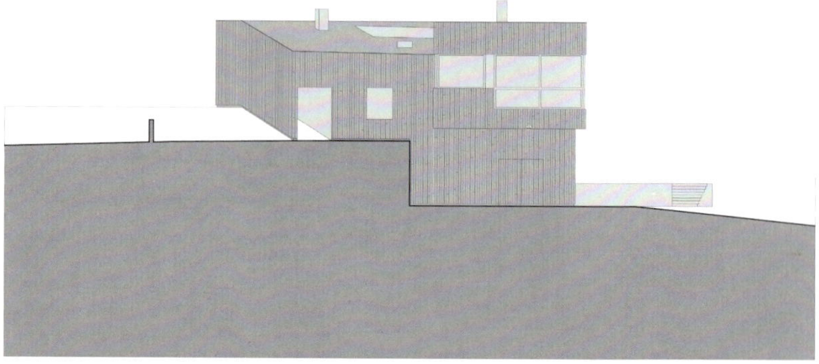

住宅K位于一处大型住宅地块，Sljeme山脚下一片森林边向东倾斜的缓坡上。目标建筑在这块地块的西北角临街而建，首先映入眼帘的就是美丽的森林景观；崎岖的地形刚好能用来拓展目标建筑的空间。

住宅设计理念的灵感来自蜗牛壳，它围绕轴心螺旋盘旋，因此，住宅的所有功能元素互相衔接，围绕中庭而建。设计中庭是为了弥补客厅区光照的不足，确保房间两边都能得到阳光。

住宅尽可能敞开，走上两层楼，就可以看到大片的绿色草地和森林；一楼简洁的外墙和隐藏的车库门封闭了住宅临街的一面。

和蜗牛壳的循序渐进相反，住宅K的入口缓和而不明显，位于南墙壁的边缘。正是从这里，围绕中庭，半层接半层的螺旋上升开始；尽头是宽敞的客厅，通向室外阳台和森林。

为了满足居住者的各种需求，住宅内设计了丰富多彩的元素，面积为600 m^2，分为几个半层，因此可以尽可能地利用地势高度差，看起来让住宅外观更小巧。

所有外墙和阳台都采用天然柚木板进行镶嵌。与螺旋渐进相应，并且为了营造亲密舒适的氛围，内部延用了木质镶板。因此地板和墙壁交错镶嵌柚木板，天花板和墙壁主要是白色。黑色，作为墙壁的辅助强调色，主要是为了引人注意，体现动态感。

养生区和游泳池巧妙地衬托了住宅内部简洁的配色。因黑色格架为主打的天花板而显得昏暗的气氛，通过从紫到蓝三色灯光变幻，营造出真实的感觉，使得此区域的背景设计具有立体感。

屋顶露台为住宅增添了奇特而圆满的亮色。第一层和屋顶间中庭里的外部螺旋楼梯一直延伸到这个露台，所以突出了这样一个事实，即，家庭住宅，除了是家人的住所之外，也是聚会、放松、享受和得到满足的地方。

House V

住宅V

Site area: 3302 m^2 Designed by: 3LHD Architects

用地面积：3302 m^2 设计：3LHD建筑事务所

House V's design and careful integration with the landscape set it apart from the context of neighboring buildings. It is situated in parallel with the street, has an elongated floor plan and takes a defensive position towards the street to which it opens only through small vertical glass windows in order to protect the inhabitants from unwanted views and noise. On the other side, the ground floor of the house opens towards a park with long horizontal sliding openings and draws nature inside.

The emphatic longitudinal plan of the house imposed an almost conceptual mastering of the house's elements. Thus the entrance is logically placed at one end next to the garage, while the swimming pool is at the other end of the home. The space flows continuously from entrance to kitchen and dining area, passes through the living room, a work area, a gym and terminates at the swimming pool. The entrance to the house is unobtrusively characterized by a volume of massive wooden paneling, which conceals within itself several basic functions of the entrance area, such as a toilet and a dressing room, as well as a library that arises as soon as one enters the living room.

The living area is the most central and highest point of the house, and it is from here that one goes to the bedrooms. This is also a position from

which the home's dimensional progression changes; it is the truest site for spatial exploitation. An almost sculptural staircase made from massive oak and white MDF opens into an airy gallery at the upper level and creates a necessary counterbalance to the unpretentious remainder of the house.

The bedroom vicinity of the gallery is easily reached through a funnel-shaped corridor that leads into three children's rooms and the parent' s consignment at the end. The interior is dominated by light tones of walls and wooden floors while the exterior of the house is a slightly darker plastered façade repeatedly punctuated by aluminum openings.

The outer terrace continues to follow the longitudinal shape of the house and is in direct contact with sliding glass outer walls on the ground floor enabling several exits to the landscape. The accentuated longitudinal design of the space is additionally heightened by a porch above the longitudinally placed terrace, which acts as both a veranda and an external living room area for the upper rooms.

The exquisite quality of the clean and vast meadow, marked on one side by the forest and the house on the other, preserves and intensifies its natural character without any additional landscaping.

住宅V的精心设计与周围景观融为一体，丝毫不受附近建筑的影响。它与街道平行，从平面看呈细长型，保守地朝向街道，为了使居住者免受不必要的景观或噪声的干扰，只有小巧的垂直玻璃窗才通向大街。另一方面，住宅的一楼通向一个公园，长长的平滑的出口，将大自然引入室内。

住宅强调长向设计着重对住宅元素的把握。因此，入口理所当然地自成一端，靠近车库，而游泳池应在住所的另一端。从入口到厨房和餐厅，空间流畅，经过客厅、工作区、健身房，末端是游泳池。住宅的入口采用大量厚重的木质镶板，简明扼要，掩盖了入口处几个基本的功能区，但一旦走进客厅，就可以看到卫生间、更衣室，以及书房。

生活区是住宅的中心和亮点，正是从这里，走进卧室，也正是从这个方向，住所的空间连续性开始发生变化；它是空间开发最真实的一例。厚橡木和白色的中纤板做成的雕刻楼梯通向楼上的通风长廊，巧妙地弥补了住宅其他部分的含蓄。

卧室靠近通风廊，通过漏斗形的过道到达三个儿童房以及尽头的父母卧室，非常方便。内部墙壁以浅色和木质地板为主打，而住宅的外部则略显暗沉，用几扇铝合金大门进行点缀。

外部的阳台沿用住宅细长的形状，直接和一楼的滑动玻璃外墙相接，使得在各处都能观赏到风景。空间强调纵向设计，这一点通过纵向阳台上的门廊表现得更加显著，它既是一个走廊，又是楼上房间的外部客厅。

大片清洁的草地，优美而细腻，一边连着森林，一边连着住宅，具有自然特色，任何其他的风景都显得多余。

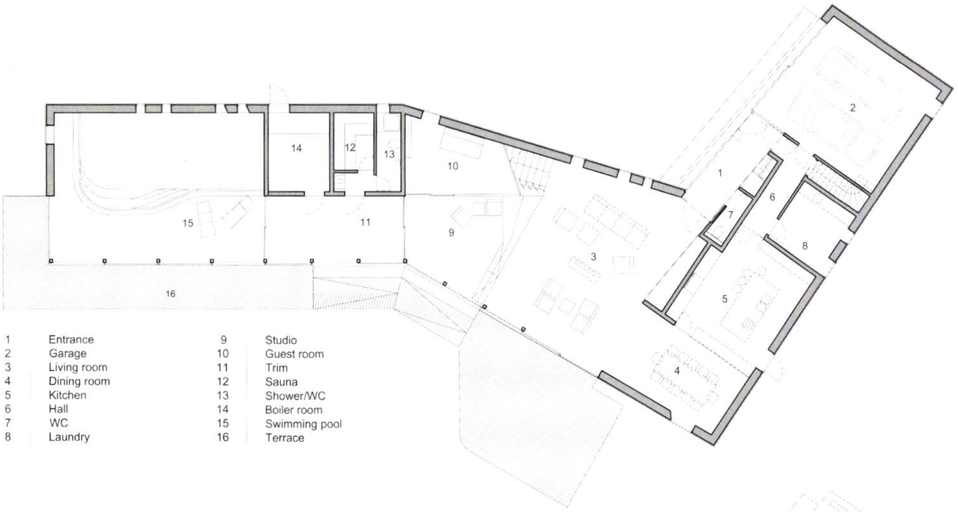

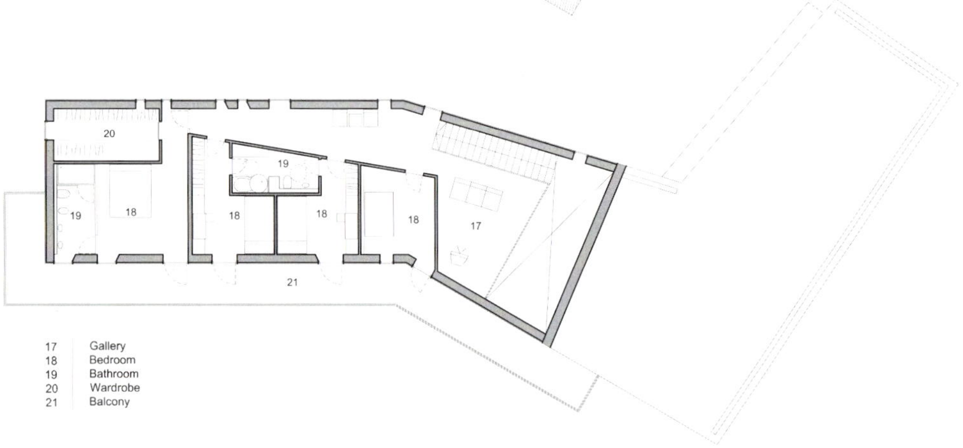

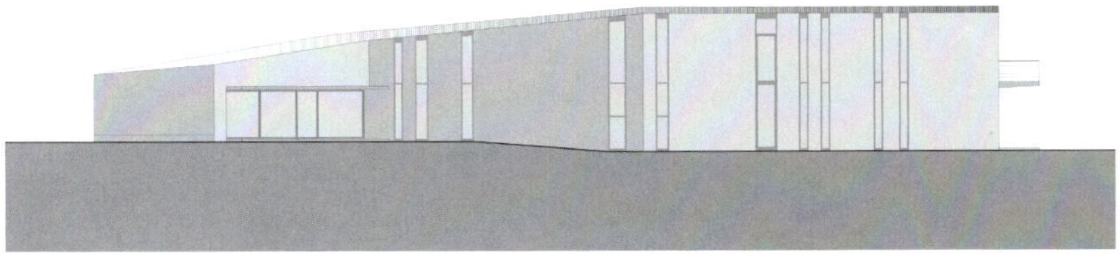

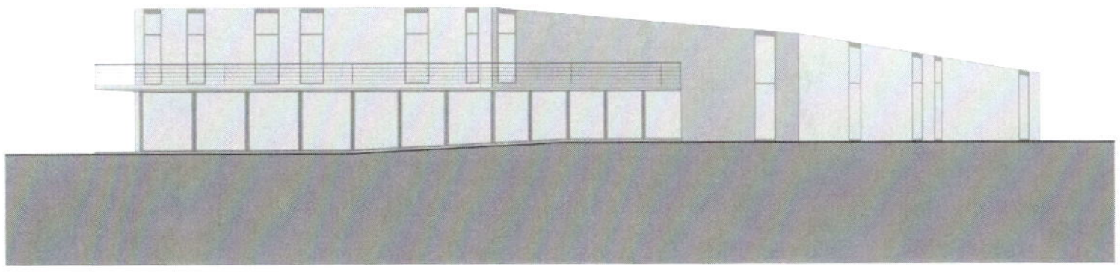

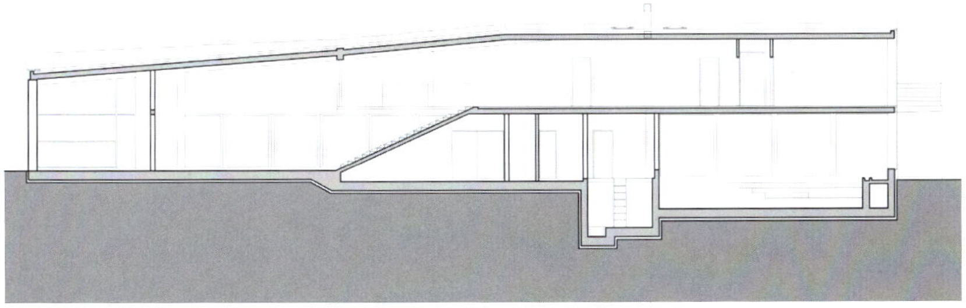

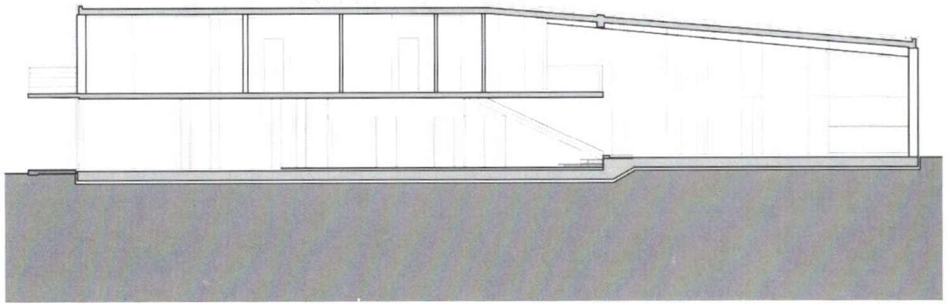

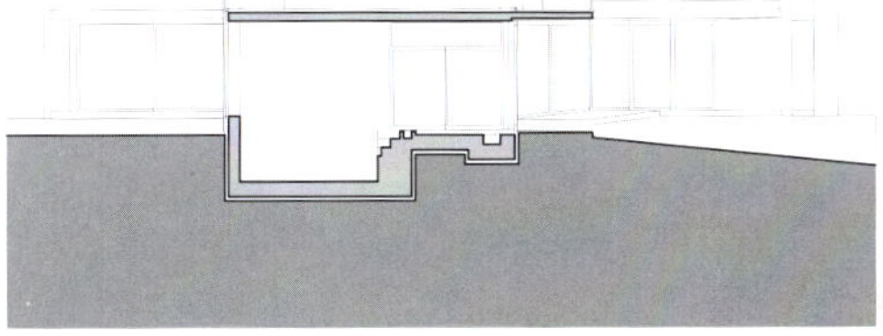

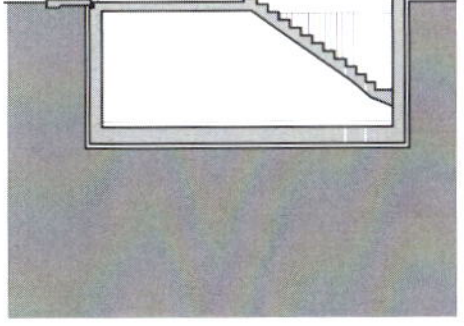

Waterfront Villa Residence, Tianjin

天津梅江南水岸公馆别墅

Building coverage: 20000 m² Architectural designed by: Allied Architects International (AAI)

建筑面积：20000 m² 建筑设计：加拿大AAI建筑设计事务所

To meet people's requirements for waterscapes and outing in spring, Waterfront Villa Residence connects the central green land with the waterfront platform by a green corridor. Thus all the landscapes connect with each other closely, open or closed with corridor or shading, to provide people with different feelings. The design integrates people, architectures, water and green land together to form multi-layer landscapes with the lake in the center, waters running and green walkways between.

The entrance hall is spacious and bright with two corridors leading to the living room which also connect with the garden through the outdoor corridor to create a diversified space for the inhabitants. All the master bedrooms are designed with a study room, a locker room, a washroom, a terrace and so on. The personalized workroom and recreation room connect with the sunken garden which connects with the outdoor main courtyard directly. The flexible space is well-arranged to achieve the natural transition between the public space, semi-open space and the private space.

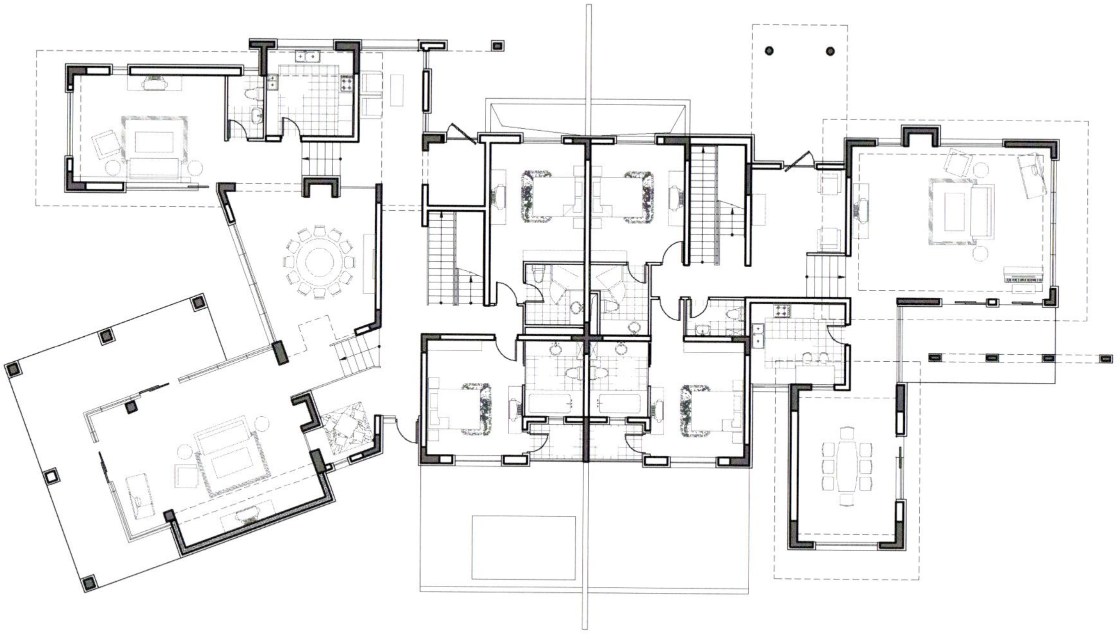

水岸公馆为实现居住者观水与踏青的双重享受，将组团的中心绿地用一条绿色走廊与岸边亲水平台相接，环境景观环环相扣，有开阔、有封闭，有通廊、有遮挡，景致不同而耐人寻味。将人、建筑、水、绿地融为一体，形成以湖面为中心、河流相伴随、绿化步行系统相衔接的多重景观视觉，体现了对居住者景观视线的关怀。

别墅入户的门庭宽敞明亮，两层通高的通廊直接通向起居室，越过外廊与花园相接，为居住者创造了丰富的空间意境。每户主卧室均有书房、更衣间、盥洗室、露台等服务空间，富有个人情趣的工作室、娱乐室与下沉庭院相通，可直达户外主庭院。灵动空间分区动静合理，体现从公共区域、半公共半私密空间到私密空间的自然过渡。

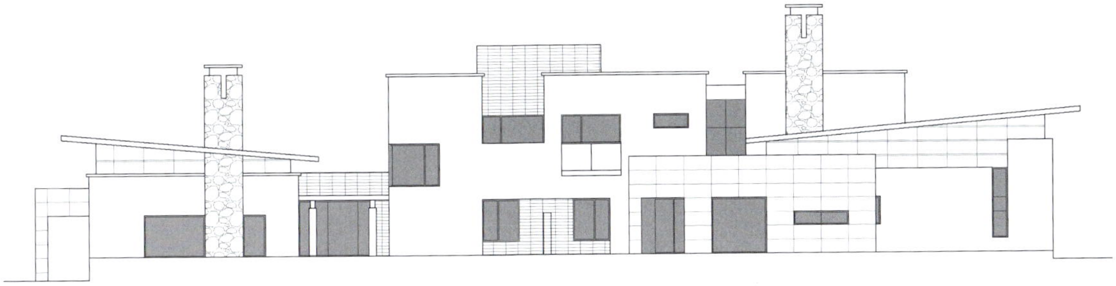

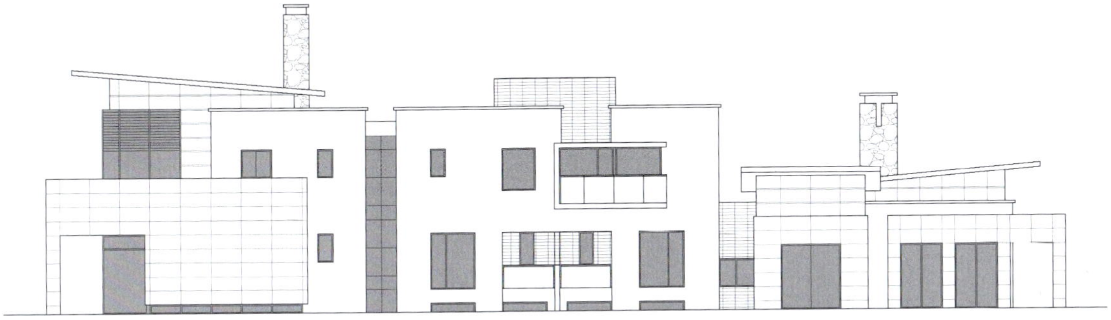

The Waterfall House

瀑布之家

At the back of the lot, at the ground level the lake has a shy presence. So it was considered to raise up to get a better look at the lake. This is the reason why it was decided to place the social area in the upper level, conquering the panoramic views to the lake during the day.

It is there where the water shows up. Unexpected at that height, it embraces the architecture and paints it with reflections, bringing life and movement. Appearing and disappearing to the eyes of the user, relaxing with it's sound, dazzling with reflects, going through a path that stimulates the senses.

The shallow pool expands the house limits, making them infinite. It erases the line between natural and artificial, inside and outside. The elevated water falls, as a waterfall, over the swimming pool in the lower floor, refreshing the rooms and providing privacy at the same time. The water fall also filters the northern light making it less harmful.

The waterfall is the icon of this house; this is why it was named after it.

About the functionality of the house, flexible spaces are designed, so it could be adapted as much as to the social active life of our client, as to the needs of a future family. The flexibility allows isolating the private areas in the lower floor from the public areas in the first floor and part of the lower floor.

As a way to connect the street life with the inside of the house, was designed a mixed use room, used as a garage, playroom or to host parties. The vertical circulation in triple height baths the inside with cenital light, joining the different levels.

在该建筑的后面，有一个湖泊若隐若现。因此，要抬高视角以更好地观察湖泊，这就是为什么决定把社交会所放在上层的原因，这样白天才能将湖光景色尽收眼底。

正是因为水的出现，在这样的高度，湖泊拥抱着建筑，湖光倒影渲染整个建筑，带来了生机与活力。湖光倒影若隐若现，令人放松的声音，波光闪闪的湖面，穿过这样一条小路，令人心情愉悦。

浅池扩大房子的范围，变有限为无限，天然与人工，内部与外部的界线消失了。较低楼层游泳池上的高架水瀑布令房间清凉，同时也提供了一些隐秘性。瀑布也挡住了一些光线，减小了光线的危害。

瀑布是这套房子的标志，这就是以它命名的原因。

关于房子的功能，空间设计灵活，因此它可以尽可能调整到适应客户活跃的社交生活，也可以适应未来家庭的需要。这种灵活性可以把较低楼层的私人空间和一楼的公共空间与较低层的部分空间分隔开来。

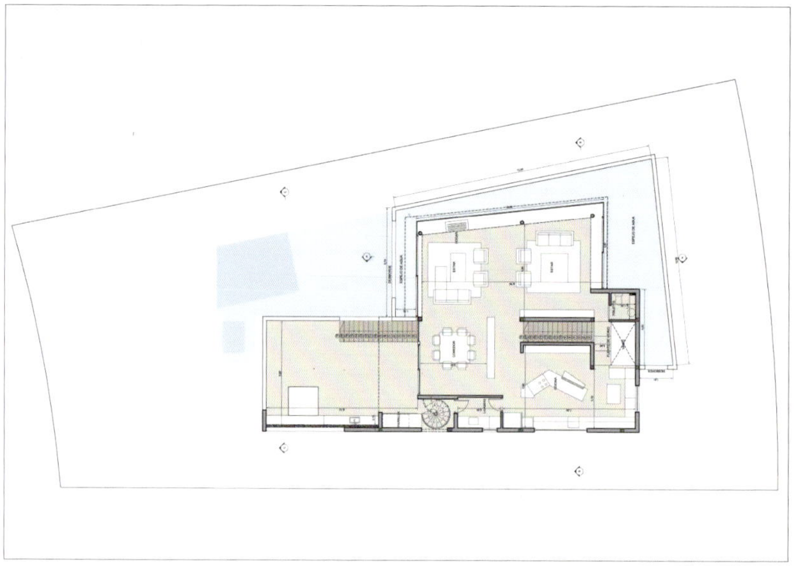

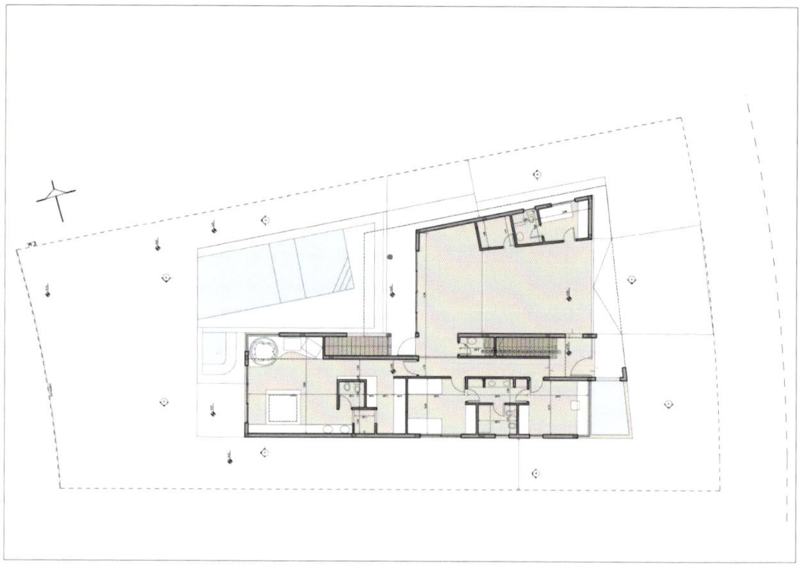

+9.50m
CONDUCTO DE CHAPA
+8.75m
+8.25m
REVESTIMIENTO MONOCAPA TARQUINI COLOR GRIS
REVESTIMIENTO MONOCAPA TARQUINI COLOR GRIS
+5.56m
CARPINTERIA DE ALUMINIO
+5.14m
+3.74m
+3.62m
REVESTIMIENTO MONOCAPA TARQUINI COLOR GRIS
PORTÓN REVESTIDO EN MADERA
+1.00m
+0.00m N.P.T. P.B.

+8.75m
+8.25m
REVESTIMIENTO MONOCAPA TARQUINI COLOR GRIS
+6.36m
REVESTIMIENTO MONOCAPA TARQUINI COLOR GRIS
+5.56m
CARPINTERIA DE ALUMINIO
+3.62m
+3.14m N.P.T. P.A.
+3.14m N.P.T. P.A.
REVESTIMIENTO MONOCAPA TARQUINI COLOR GRIS
+2.00m
+1.00m
REVESTIMIENTO MONOCAPA TARQUINI COLOR GRIS

+8.75m
REVESTIMIENTO MONOCAPA TARQUINI COLOR GRIS
+8.25m
+6.36m
REVESTIMIENTO MONOCAPA TARQUINI COLOR GRIS
CARPINTERIA DE ALUMINIO
+5.56m
REVESTIMIENTO MONOCAPA TARQUINI COLOR GRIS
PORTÓN REVESTIDO EN MADERA
CARPINTERIA DE ALUMINIO

+8.75m
REVESTIMIENTO MONOCAPA TARQUINI COLOR GRIS
+6.36m
REVESTIMIENTO MONOCAPA TARQUINI COLOR GRIS
+5.56m
REVESTIMIENTO MONOCAPA TARQUINI COLOR GRIS
REVESTIMIENTO MONOCAPA TARQUINI COLOR GRIS
CARPINTERIA DE ALUMINIO

The Black House

黑屋子

Surface: 330 m² Design team: andres remy, hernan pardillos, julieta rafel, carlos arellano, gisela colombo

占地面积：330 m² 设计团队：安德列斯·雷米、埃尔南·帕蒂罗斯，胡列塔·拉斐尔、卡洛斯·阿雷拉诺、吉塞拉·科伦坡

This single house is located in a closed neighborhood, 30km. away from Buenos Aires. The lot, 20 meters wide and 50 meters length with 3 meter of lateral retreats, has amazing views to the lake we could not let aside.

The best views to the lake were at the back of the lot, while the best orientation was at the front. The shallow pool that divides the house in two allows the indirect light to bathe the interiors, as the northern sun reflects it's light on the water surface. This way, light is present in every corner of the house, but never in a direct way.

Both programs are connected by a glass bridge, with the water running under your feet. The living room, 10 meters wide and 5 meters length, opens to the exterior using glass walls. It was thought in a lower level than the rest of the house, making it permeable and allowing the ambients a clean view to the lake.

The resolution of the first floor follows the same criteria of differencing areas. At the front are located the bedrooms for the future children, with views to the lake. As a bridge, joining the two volumes in the lower floor is the main bedroom with a giant overhang that conquers the best views to the lake, seeming to float over the water.

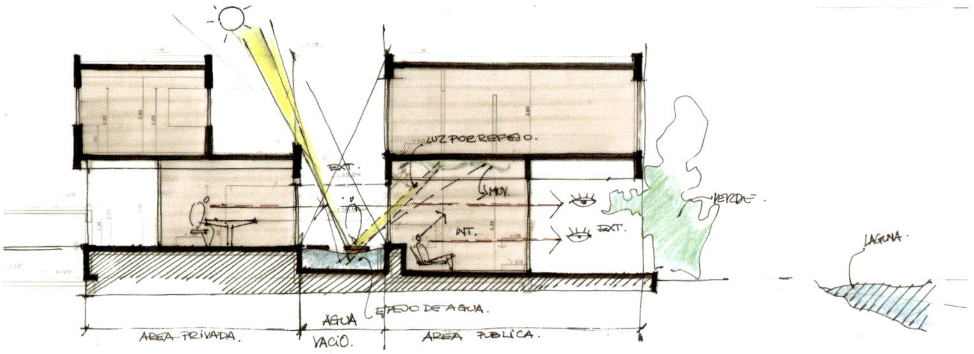

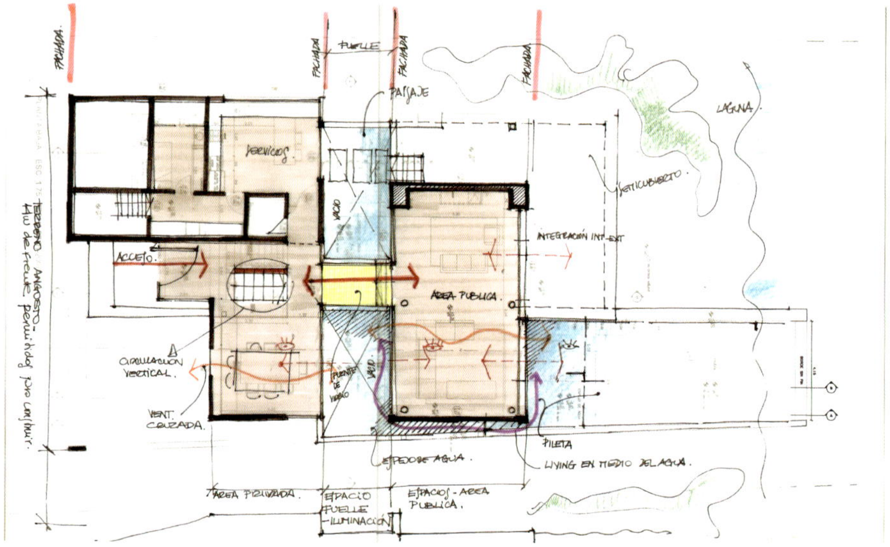

The Black House has an almost provocative sobriety, where the pure white in the inside provoques an emotive contrast with the absolute black in the outside, reminding the bite of an apple. A strong characteristic that names the house.

这种独户房子坐落在一个封闭的社区，距布宜诺斯艾利斯30km。该房子占地宽20m，长50m，侧面向后倾斜3m，能观赏到迷人的湖泊景色。

观赏湖泊的最佳视角在房子的后面，而最佳取向在前面。浅池将房子一分为二，北方太阳的光线反射在湖面上，房子内部沐浴在间接光线中。这样，光线充满房子的每一个角落，但所有光线都是间接的。

这两个项目由玻璃桥连接，脚下湖水波动。客厅宽10m，长5m，透过玻璃幕墙与外部相通。客厅的设计比房子内其他空间要低一些，使客厅更具通透性，湖面景观更简洁明快。

一楼房间的拆分符合区分房间的标准。前面是为未来孩子准备的卧室，带有湖面景观。连接较低层两个建筑群的是主卧室，作为两个建筑群间的桥梁，带有一个巨大的悬垂部分，像是漂浮在水面上，站于此，可将湖面最佳景观尽收眼底。

The Orchid

兰花楼

Land area: 3640 m²

Designed by: Andres Remy Arquitectos

占地面积：3640 m²

设计公司：安德烈斯·雷米建筑师事务所

The Orchid is a sustainable architecture. Sustainability implies a lot of varieties, such as efficient and rational use of energy and water, natural ventilation and lighting, and low-environment-impact materials. The house has the best orientation, possible thanks to the large lot.

The concept came from the owner's hobby, growing orchids. The house is based on the different parts of the orchid: the roots, the stem and the flower.

The sun rays impact in the interior of each room, were also studied, to determine the optimal depth to place the windows. This gives a unique volumetric outcome to the project. Taking advantage of sun rays in winter increases the interior temperature up to a comfortable level. The design includes glazed volumes with good thermal insulation and small windows in the worst orientation, such as the south facade.

The windows consist of aluminum frames supplied with thermal bridge breaker and double-hermetic-glass. A wide variety of insulating materials were also used, as well as water based paint and wall and roof air chambers.

In the lower floor, the convenient location of the opening windows allows

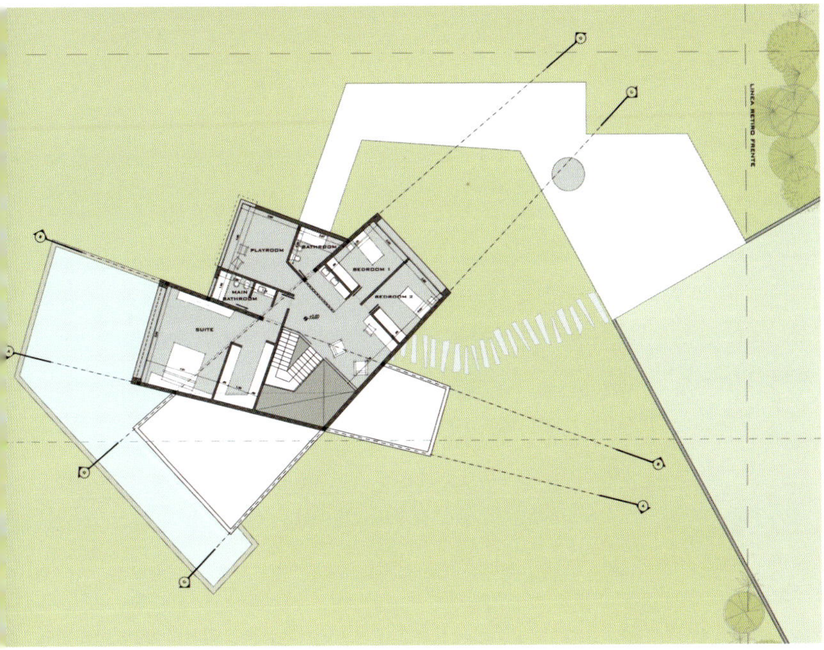
PLAYROOM
BATHROOM
BEDROOM 1
MAIN BATHROOM
BEDROOM 2
SUITE
LINEA RETIRO FRENTE

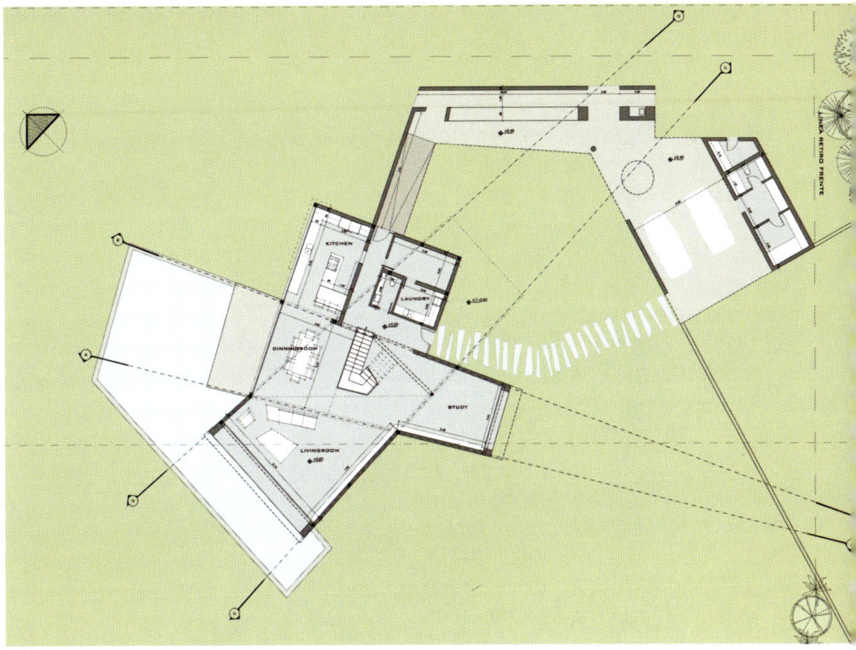
KITCHEN
LAUNDRY
DINNINGROOM
STUDY
LIVINGROOM
LINEA RETIRO FRENTE

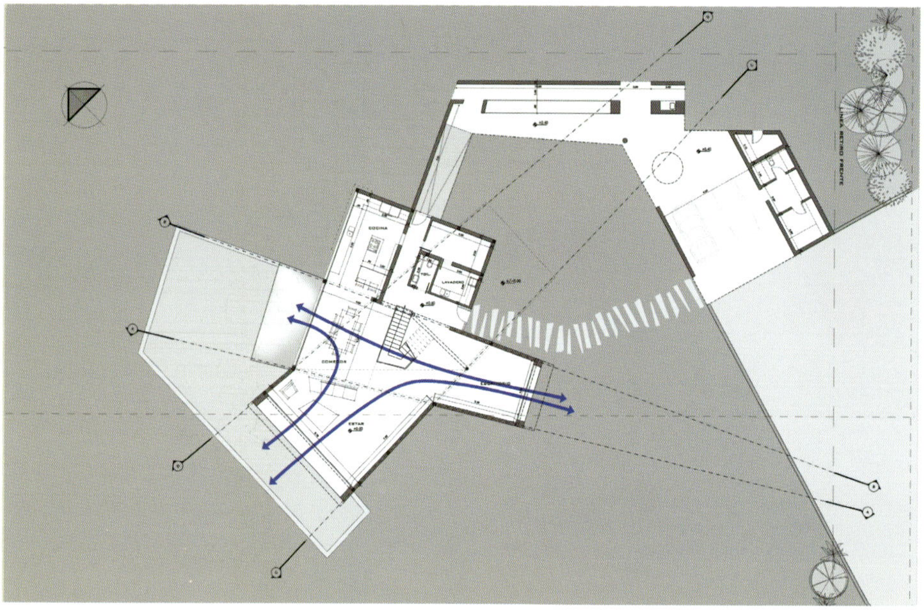

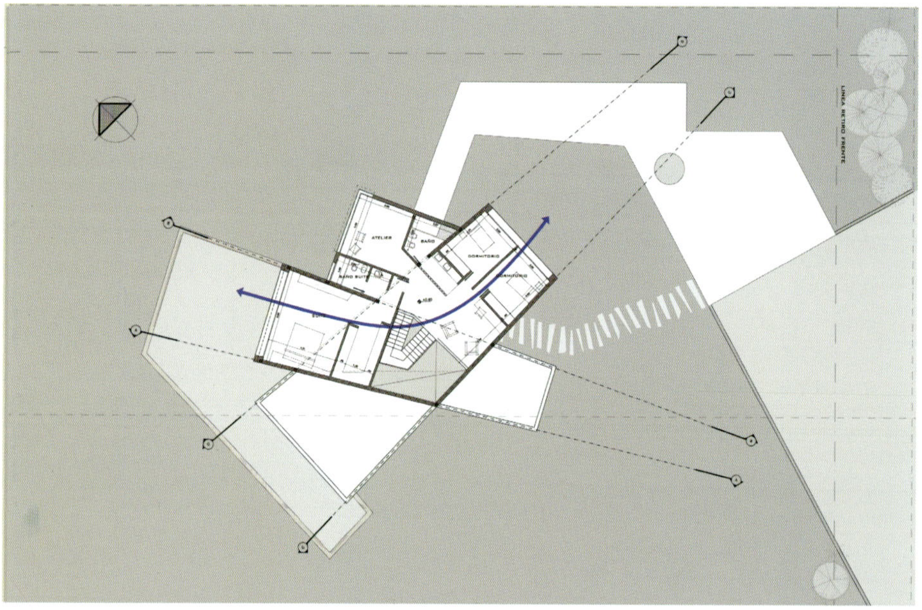

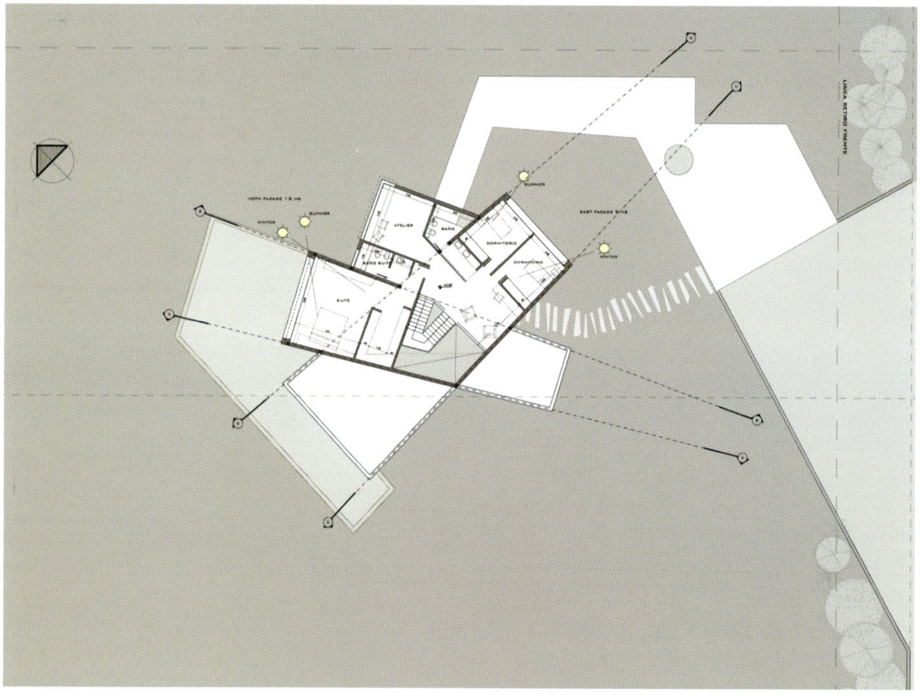

natural air flow helping to decrease the humidity in the room. What's more, the big thresholds create a good distribution of fresh air.

All this, gives account of the complexity of the house, a glass and concrete flower, designed according to a program of needs and the owner's concerns.

兰花楼是一种可持续的建筑，可持续性意味着使用的品种很多，如高效合理的利用能源和水，自然通风和照明和利用对环境影响低的材料。这房子有最好的方向，很可能要归功于巨大的面积。
这一建筑概念来自于业主的业余爱好：栽种兰花。房子基于兰花的不同部位：根、茎和花。

兰花楼研究了太阳光线对每个房间内部的影响，以确定安装窗户的最佳深度。这使该项目有了一个独特的体积感。冬季利用太阳光增加室内温度，使之更加舒适。该设计在最差劲的方向，比如南立面，使用了带良好热绝缘的釉面和小窗口。

窗户由铝框和热桥断路器与双密封玻璃构成。也应用了各种各样的绝缘材料，还应用了水基涂料以及墙体和屋面气室。

在较低的楼层，开窗使自然空气流通，有助于减少室内的湿度。另外，大门也引进了大量的新鲜空气。

房子是一朵玻璃和混凝土构成的花，是根据项目的需要和业主的要求设计的，所有这一切都考虑了房子的复杂性。

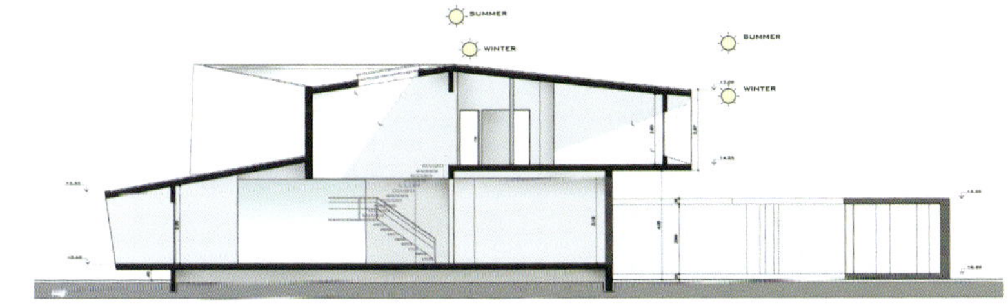
SUMMER
WINTER
SUMMER
WINTER

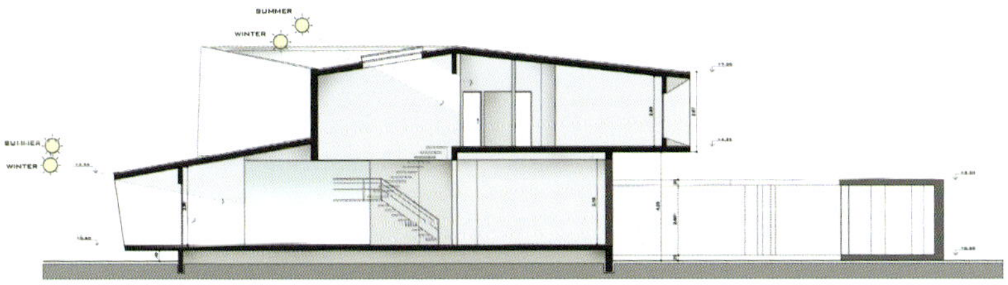
SUMMER
WINTER
SUMMER
WINTER

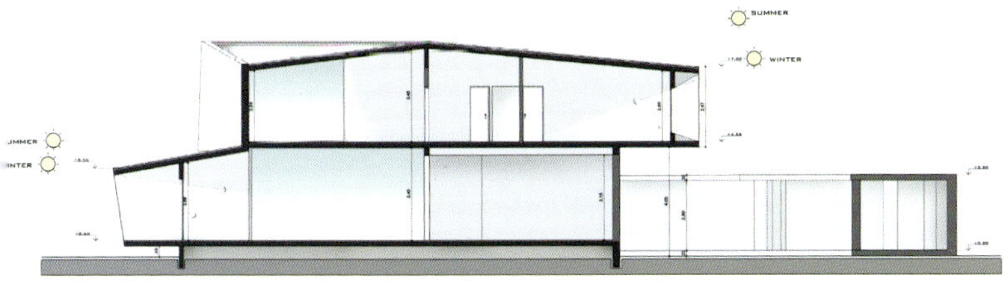
SUMMER
WINTER
WINTER

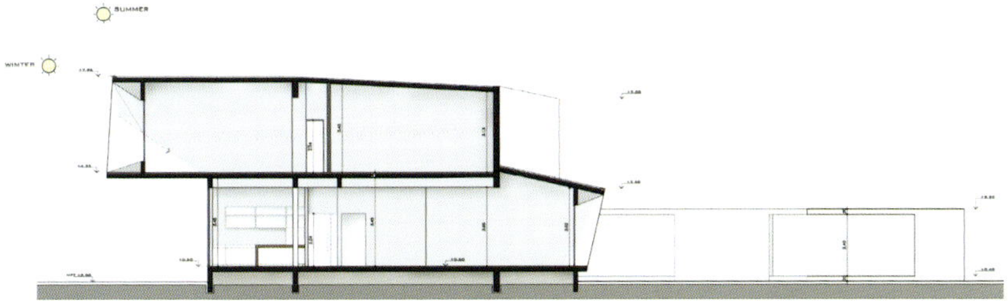
SUMMER
WINTER

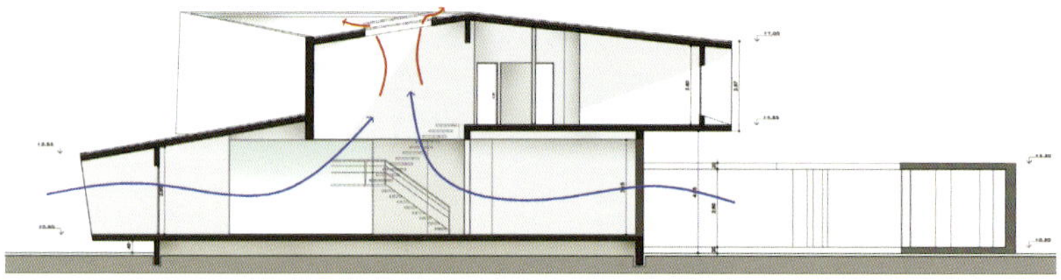
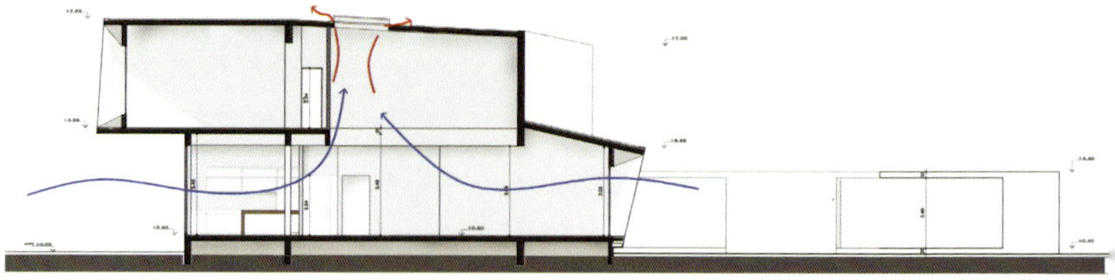

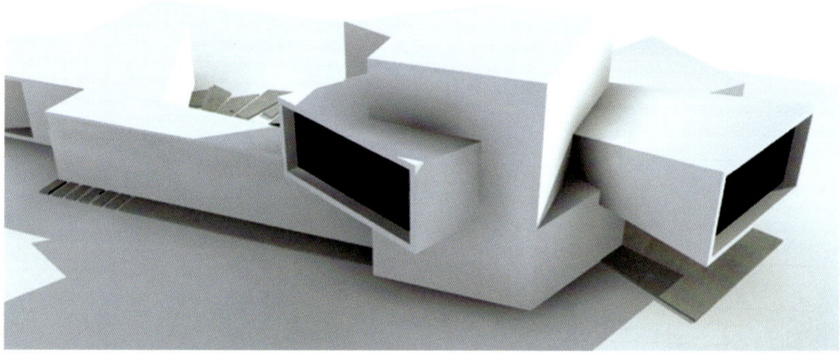

RZ House

RZ住宅

Area: $270m^2$ Address: Santa Barbara, Benavidez, Provincia de Buenos Aires

面积：$270\ m^2$ 地址：圣巴巴拉，贝纳维德斯，布宜诺斯艾利斯省

The design of the house proposes the connection of two volumes as a response to the site´s special landscape conditions. At the crossing of the canal and the lake, the house captures the best views from the main rooms while the more private areas stay less exposed. The plan is a J letter which adapts to the site´s conditions and with a 30 degree angle creates the main entrance to the house.

At the rear of the building, the windows frame different views of the outstanding surrounding landscape while at the front, the openings are horizontal and being set at the viewer´s height allow the continuous view of the lake.

The relationship between the swimming pool and the house is a very special one, in which the former captures the sunlight by reflection creating an appropriate microclimate for all the rooms surrounding it.

房子的设计将两个建筑群连接在一起，来呼应该地的特殊景观条件。在运河和湖泊交叉处，从主要用房可以获得最佳景观，更私密的空间则与外界接触较少。该计划是一个“J”字形，与该地的条件相适应，30°角的拐弯创造了房子的主入口。

建筑物后方，窗户形成了周围美丽景观的不同框景，而在前方，水平的开口和观望者的视线高度持平，可以看到一望无际的湖泊。

游泳池和房子之间的的关系非常独特，游泳池反射太阳光，为周围房间创造一个温暖适宜的小环境。

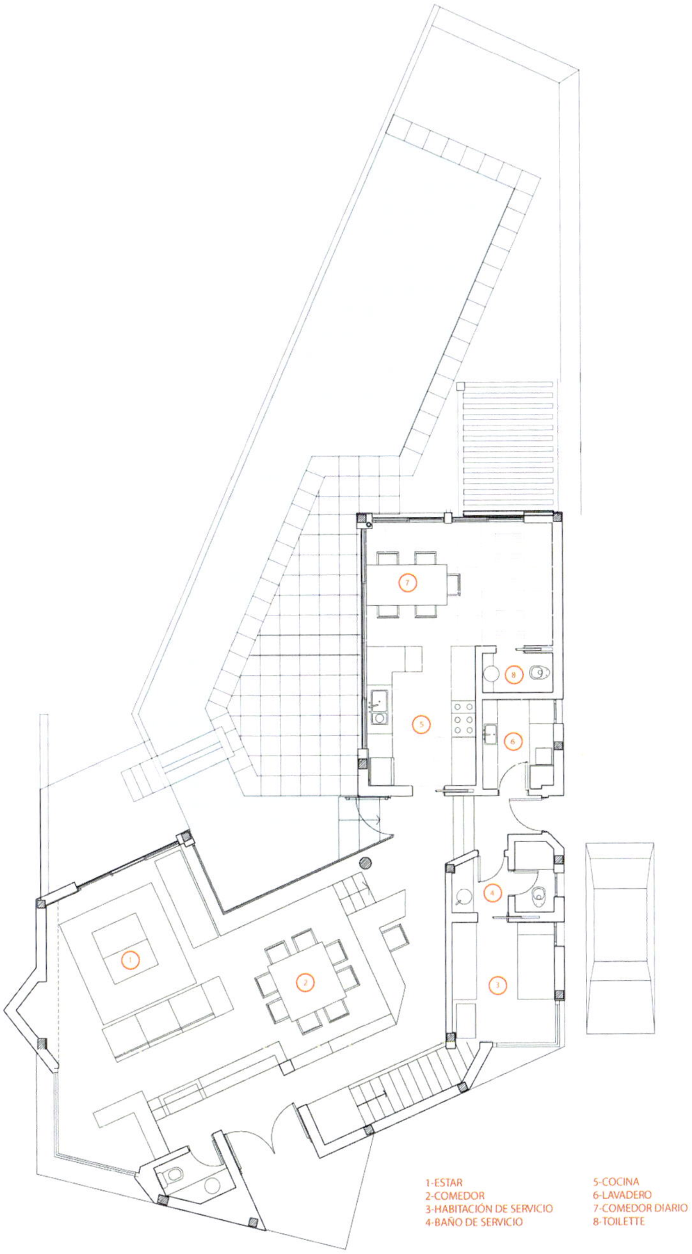

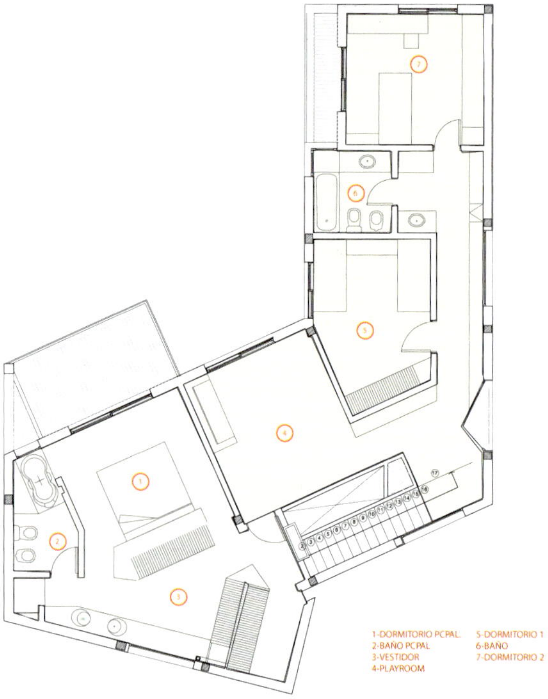
1-DORMITORIO PCPAL.
2-BAÑO PCPAL
3-VESTIDOR
4-PLAYROOM
5-DORMITORIO 1
6-BAÑO
7-DORMITORIO 2

RM House

RM住宅

Site: 760 m²　Location: Buenos Aires

总平面面积：760 m²　位置：布宜诺斯·艾利斯

Situated in "Laguna del Sol", in the outskirts or Buenos Aires, the RM house raises in a 760m² lot.

Designed for a young couple and its two kids, with an active social life, Andrés Remy came out with a house with flexible, cozy spaces maintaining clean lines.

The client's premise was to use up to 300m² of the lot, allowing the green area to embrace the house. This creates a fluent connection with the outside without giving up privacy.

The lower floor contains the public area, which is divided throughout different levels as well as changing flooring materials like cherry wood and polished cement.

An indoor green area in double height lets light fill the interior, and articulates the circulation at the same time.

A sculpture stair leads to the playroom in a middle level.

In the upper floor, the green area splits the functions in two. For one side

is the main bedroom, for the other side the kids bedrooms.

A modern and functional house, with the poetical strength to affect the life of its inhabitants by means of small details that make the difference.

设计师安德烈·雷米推出了一款新房子，空间灵活舒适，线条简洁，专为有活跃社交生活的年轻夫妇和他们的两个孩子设计。

房子的房屋和地基面积高达总面积的300m²，房子得以有足够的绿化面积。这使房子与外界有充分的联系，而又不失隐秘性。

下层包含了公共区域，公共区域分为不同级别，且应用不同地板材料，如樱桃木和抛光水泥。

双高房间的室内绿化面积使室内光线充足，同时令室内空气循环良好。

在中层由雕塑楼梯可到达娱乐室。

在上层，绿化面积把房间分为两边。一边是主卧室，另一边是孩子们的卧室。

这是一个现代化的多功能住宅，细节处都有些许差异，诗意般的境界影响着居住者的生活。

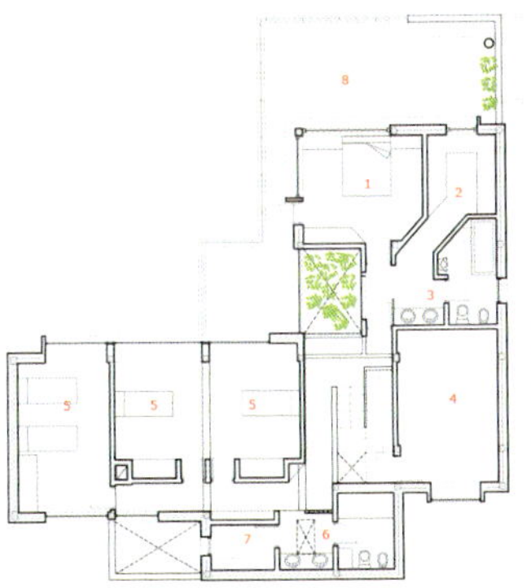
PLANTA ALTA
1. Dormitorio Principal
2. Vestidor
3. Baño Principal
4. Playroom
5. Dormitorio
6. Baño
7. Depósito
8. Terraza

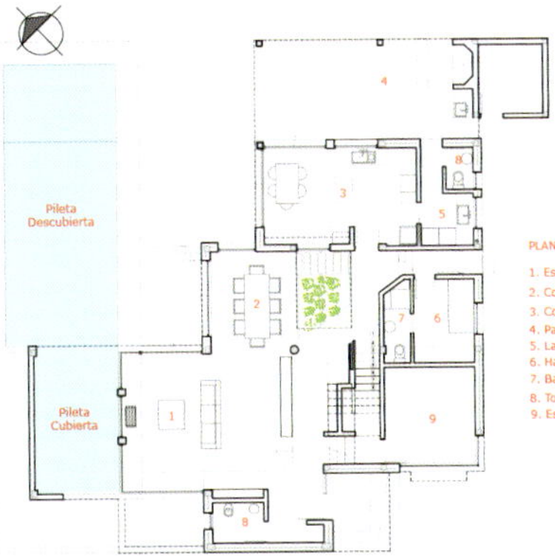
Pileta Descubierta
Pileta Cubierta
PLANTA BAJA
1. Estar
2. Comedor
3. Cocina Comedor
4. Parrilla
5. Lavadero
6. Hab. de Servicio
7. Baño de Servicio
8. Toilette
9. Escritorio

SHELL

贝壳

Total floor area: 329.65 m^2

总建筑面积：329.65m^2

Architects: Kotaro Ide / ARTechnic architects

建筑师：井孝太郎的IDE / AR工艺学建筑师

A large shell shaped structure finds itself in the middle of the woods. It is hard to determine what exactly the structure is, and unlike the surrounding caves and rocks, it clearly is not a part of nature – nor is it a ruin. A frame, a shape, made at a completely different place for a completely different purpose. Within this shell shaped structure will one find floors constructed, wall separating spaces, and rooms furnished. The scenery conjures a SF film-like image, in which locals inhabit over an abandoned spacecraft. With time, trees start to grow encircling the spacecraft, harmonizing it into the landscape.

By isolating living space from the wilderness, and upgrading its quality as a shelter, the house will be protected from nature and will provide a comfortable environment. With this, the house will be taken care of and used frequently and continuously.

The regions' low temperatures and high humidity level makes for a harsh climate. Despite the general avoidance of concrete material in the region, its usage and the lifting structure have helped the villa protect itself from the humidity.

Leaving the boundary between human life and nature ambiguous is a Japanese virtue. Yet, this ideal can only be achieved through meticulous

attention and care of the wilderness on a daily basis. In the style of many modern sculptures, the designers aimed to enhance the surrounding nature by incorporating it within the spatial structure.

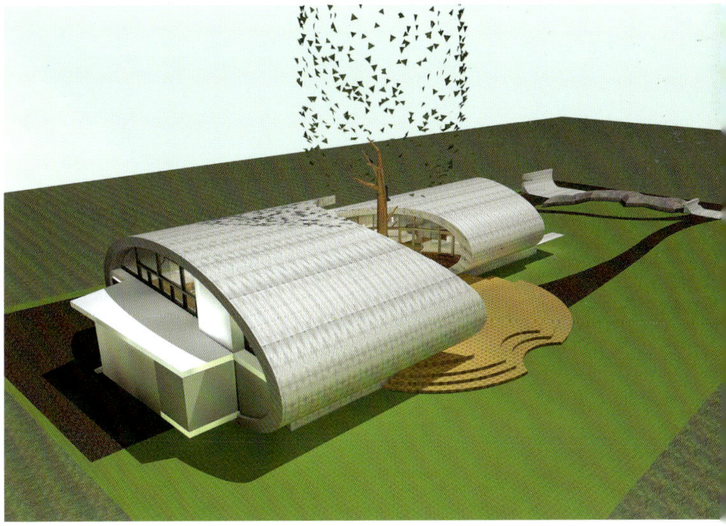

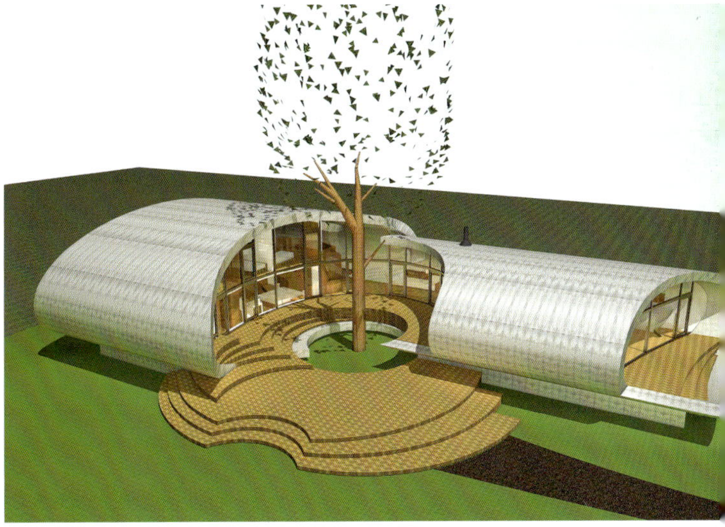

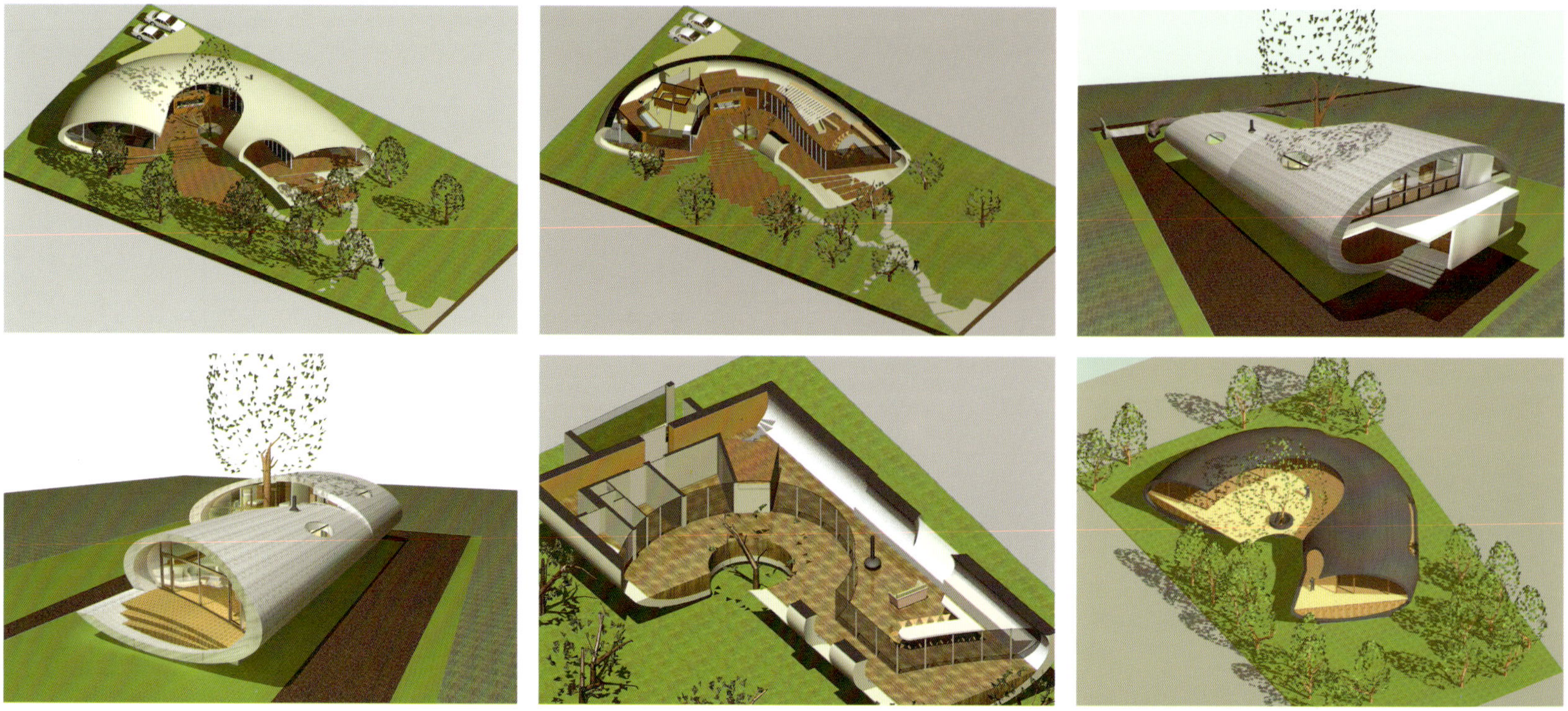

丛林中有一个大贝壳状的结构。很难确定这个结构到底是什么，显然不像周围的山洞和岩石，他们属于自然的一部分，也不是废墟。它是在独特的地方为独特的目的而设计的框架或形状。在这个贝壳状的结构里人们会发现建好的地板、有墙分隔的空间和装修好的房间。这里的景色让人想起科幻电影般的形象，当地人居住在一个废弃的宇宙飞船里面。随着时间的推移，飞船周围的树木开始生长，飞船成为景观中的和谐一景。

将居住空间与旷野分割开来，提升了住房质量，房子将不受外部的影响，并提供一个舒适的环境。鉴于此，有人将会经常照料和使用房子。

该地区有低温和高湿的特点，气候恶劣。尽管该地区一般不用混凝土材料，但使用混凝土材料和提升结构有助于别墅驱湿。

模糊人类与自然的界线是日本人的一种特质。然而，实现这一理想的途径只有仔细关心自然。许多现代雕塑风格中，设计师通过把周围的自然与空间结构融为一体，其目的是增强周围的自然美。

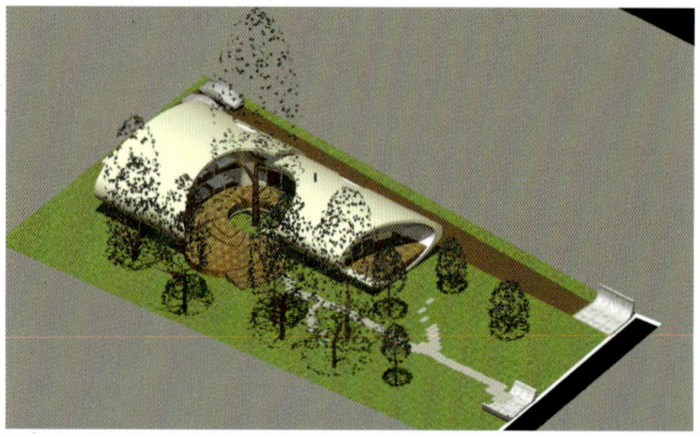

Maison du beton

杜贝特楼

Building area: 236 m^2 Architects: atelier st, Silvia Schellenberg-Thaut, Sebastian Thaut

建筑面积：236m^2 建筑设计：st工作室 Silvia Schellenberg–Thaut, Sebastian Thaut

High over a river meadow a sharp edged dwelling house rises on the gentle ridge of the Cainsdorfer local situation. It uses the available hillside situation of a generous property, contrasts by his clear geometry, nevertheless, consciously and strikingly with the grown landscape.

Town planningly the polygonale volume adapts itself in his relatival cubic capacity of course in the context of a heterogeneous structured outskirts situation. The granulation and the scale of the mostly empty single constructions of the surroundings will transfer into the solitaire construction. By the clarity and precision of the form, the big single openings and not least by the material of the extremely smooth view concrete facade, the house changes the usual picture of the surroundings.

The building was completely established as a monolithic, ferro-concrete construction with nuclear insulation. The disguising of the building incisions in the facade was realised as a curtain wall facade with anodized aluminium boards.

The internal complexity and the different space heights are not discernible from Outside and are experienceable only with entrance of the building for the visitor. Also it behaves with the creation of the interiors. Of the hardness of the concrete above all natural materials step here, like dark

wood, bright natural stone, cream-coloured materials and against colours which emit a big warmth and security in the combination.

在当地Cainsdorfer绵延的山脊上，一座棱角分明的住宅跃然矗立于河畔草地。优雅宽容的山坡，衬托出这座住宅的造型清晰明朗，虽然如此，茂密的草地依然引人注目。

城镇规划要求立体空间的多边建筑造型要与市郊多元建筑构造相适应。周围大量粗糙，以及空洞单一的建筑变得暗淡无光。外形清晰精致，宽敞的大门，尤其是看起来非常细腻的水泥外墙，这座房屋在周围建筑物中鹤立鸡群。

这一建筑十分庞大，钢筋混凝土架构，能够隔离辐射。建筑正面的开口用铝合金板材幕墙加以掩饰。

从外面并看不出内部错综复杂，只有置身其中，才能亲身体会空间参差交错。这是内部创意的体现。耐用的水泥，尤其是天然的材质，如深色的木纹，明亮的石材，米色的原料，鲜艳的对比色，相互融合，散发出温暖而安心的迷人气息。

Woodhouse, forestview

伍德豪斯森林美景

Total building area: 170 m^2

PArchitects: atelier st Silvia Schellenberg- Thaut Sebastian Thaut

总建筑面积：170m^2

建筑设计：（圣·工作室）西尔维娅·施伦贝格–萨沃特，塞巴斯蒂安·萨沃特

The rectangular and narrow construction body is cut in the ground floor and is bent. This incision forms a protected, roofed input area and separates at the same time internal functional areas in his crease points.

With this interpretation of the editions arises, with the running around same eaves high and a suitable building width, an optimum exposure of the southern roof surfaces.

Thoroughly on the rural surroundings and the nearness to the nature the dwelling house, including the roof is wrapped up, totally in fine larch timber profiles. The structure and surface of the wood changes optically, according to weather and season. In the course of the time the facade of beige, brown, silver-grey about itself will change sometime totally to a dark grey whole appearance. To all colour default of the land-use plan are carried therefore calculation.

In the teamwork between surface-terse, vaporized solar protection glasses and deep facade incisions an atmospheric facade rich in tension originates thus between lifestyle, ecology and permanence.

The ascetic and on the essential things diminished also finds in the inside area his continuation. Beside a reserved creation mediates the inside,

nevertheless, above all security and warmth.

The complicated spatial relations which particularly are effective by probably placed openings to the outside space originate from the deliberate change of differently high rooms, cut galleries and low functional areas.

A wing-smoothed, yellow screed ground underlines at all levels, as a homogeneous floor material the special spatial atmosphere.

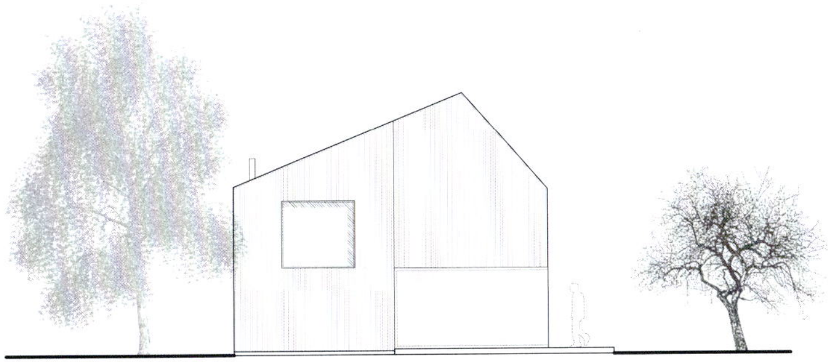

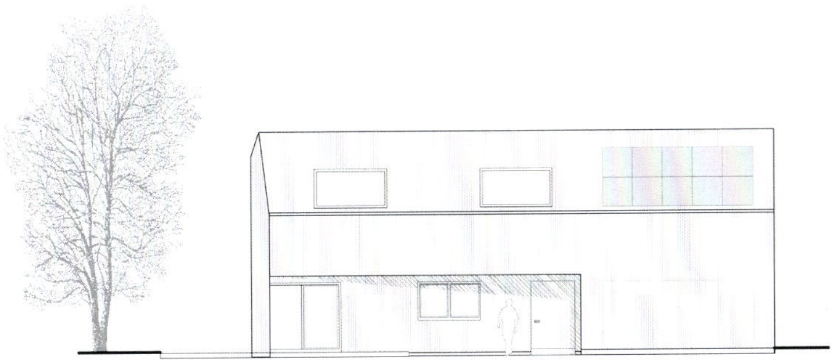

建筑体像狭窄的矩形被切割后立在地面上。这个切口形成了受保护的带屋顶的内部空间，同时折痕点把内部功能区分离开来。

根据这种阐释，高高的屋檐，适度的建筑宽度，南方屋顶的表面也能显现出来。

完全处于乡村环境中，亲近大自然，住宅和屋顶完全包括在精细落叶松木的轮廓之中。木材的结构和表面随天气和季节而变化。随着时间的变化，米色、咖啡色和银灰色的立面有时会完全变成暗灰色。计划计算了土地使用中所有颜色的变化。

表面简洁的蒸发太阳能防护玻璃和深度表面雕刻结合，在生活方式、生态平衡和显现出富有张力的美丽外观。

克己风格与消失的必要事物在内部空间得到了延续。无论如何，保守的创新调和了内部空间，超越了安全感和温暖感。

复杂的空间关系在可能与外部连接的位置尤其有效果，这种空间关系源于不同的高大房间、裁剪的画廊和低矮的功能区之间的有意变化。

不同水平的地面一律铺了表面平滑的黄色地板，这种质地均匀的地板勾勒出了特殊的空间氛围。

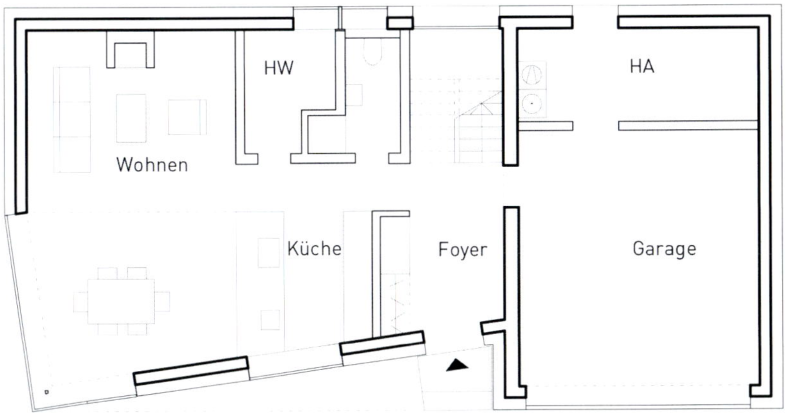

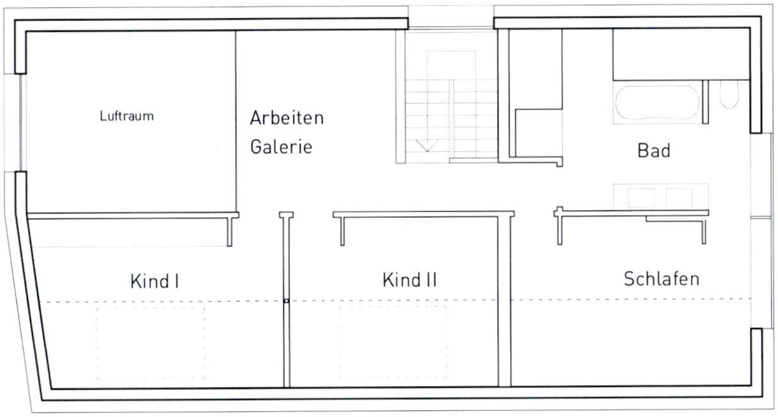

Balaam Residence

巴兰居

Gross floor area: 900 m² | Landscape: JW Concepts/ Arkhefield | Interior Designer : Arkhefield

总建筑面积：900m² | 景观：JW Concepts/ Arkhefield | 内部设计：Arkhefield

The house presents a heavily articulated and screened façade to the north and west while "peeling" back to the city views over the river to the south. Solid, playful zinc walled elements juxtapose the light weathering timber "follies" which serve different programmatic needs within the house. The clearly extroverted pool conspicuously cantilevered over the entry is the centre piece for life, play and drama within the house.

This house animates the street, the river boardwalk (both with form and sculpture) and is hopefully part of a broadening dialogue of residential architecture in Brisbane and abroad.

The form is a manifestation of the needs of the house and constraints of the site. Solid walls present to the east while the house peels back layers to the south west. In response and consultation with its neighbor, an open "garden zone" provides respite to the otherwise built up environs. The scale of the house should be considered within the broader higher density context.

住宅向北和向西是厚重的衔接和屏蔽外墙，向南“翻转”，可以观赏到河畔的城市风景。坚实而俏皮的镀锌墙体元素融合轻度风化“荒唐”的木材，体现住宅不同的主题需求。入口处惹人注目的水池是生活、娱乐和戏剧化的焦点。

住宅使街道、河流上的木板桥（形态和雕刻）都显得生气勃勃，有望成为加深布里斯班和国外居住建筑交流的一个要素。

住宅的外形体现了它的需求以及场地的限制。坚实的墙壁向东延伸，而住宅楼层向西南翻转。为了与邻近住宅相协调，开放的“花园地带”缓冲了周遭的环境。住宅的规模也应能适合人口更稠密的环境。

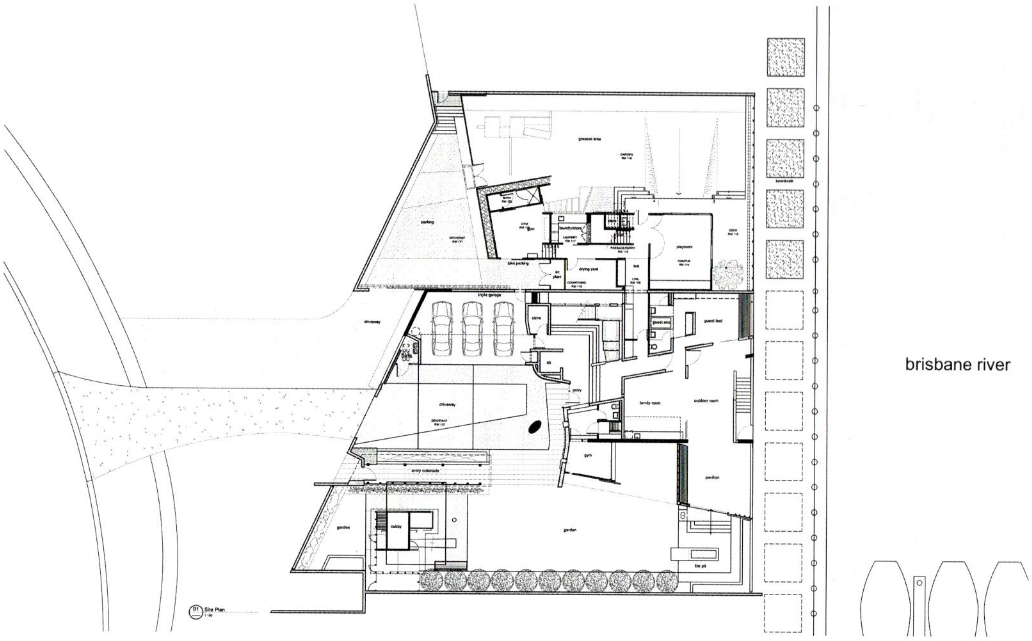
brisbane river

PRIVATE PROPERTY
NO PARKING

The V House

Designed by: Plan B Architects　Designing architects: Felipe Meza (planb) – Giancarlo Mazzanti

V楼

设计师：B计划建筑师　设计建筑师：菲利普·梅萨（B计划）– 费斯切拉·马扎提

V House is designed to enjoy a native garden and the distant view of a wetland. The designers unfold the house along the high slope of the plot, avoiding the volumetric presence in favor of a branched horizontal extension and link the deformable molecular configuration of the glass with the geometric arrangement of the house.

The molecular structure of glass, flexible and modular will define the geometric structure of the plant and the transformations that occur in the design process. The specific facts of the plot, and the particular client's lifestyle will be stresses and strains in the above structure.

Like the branches of a tree, the house falls over the site, and branches into two arms that separate the program into different heights: above, the social arm, bridge and terrace, and down the arm to private rooms and services. The main trunk is a library that starts the journey staggered from the house.

The arms are separated in two ways: the inner courtyard distances them through a native garden, and different heights allow that the zero level (ground level) of the former falls into place at roof level of the other. These two actions allow the ramifications to equally enjoy the distant view of the lake.

V楼可以享受原生花园和湿地的远景。设计师沿该地段的高坡展开房子，避免了分支横向延伸的建筑群设计，把玻璃的可变形结构与房子的几何安排联系起来。

玻璃的分子结构，灵活而又是模块化的组合，将确定该建筑的几何结构和设计过程中出现的转换。该地段的具体细节和客户的特定生活方式将成为上述结构的重点。

房子像树枝一样在该地段分布，分了两个枝干，使该项目的高度错落有致：上面是社交部分、桥梁和阳台，下面是私人房间和服务部分。主干是一个藏书室，从这里可以开始你的漫长旅程。

分支以两种方式分开：通过原生花园，内部庭院把它们分离开来，不同高度让一个分支的零水平面（地面）和另一分支的屋顶高度相同。这两种方法使两个分支可以同样欣赏美丽的湖光远景。

Zildjian
ZBT
K

CG House

CG House

Floor area: 990 m²　Designed by: GLR arquitectos

建筑面积：990m²　设计公司：GLR 设计公司

The CG House rests on a generous 1603 m² site adjacent to the Sierra Madre mountains. While the site's steep slope presents a number of design challenges. The dramatic views of the city of Monterrey provide the architect with a number of opportunities as well.

From the street, two massive oak trees rise to provide privacy and welcome shade to the terrace above the swimming pool. Due to the sloping nature of the site, a massive, exposed concrete wall defines and encloses the swimming pool and garden areas of the house and sets the tone for the design of the house itself, which is clad in brown/black volcanic stone, IPE wood, white stucco and steel.

A wide granite staircase leads to an intimate garden of palm trees. Directly ahead is a 12 feet-high dark oak door whose grand scale offers a hint of the dramatic scale of this 98941m2 house that lies beyond, beginning with an entry foyer whose 15 feet-high walls are adorned with silver leaf. A massive sculptural piece of coconut roots, steel and rocks are designed specifically for this space.

The living room, as well as the dining and family rooms, all have large windows with views of the garden, allowing light to penetrate deep within the spaces as well as creating a seamless interplay between the interior and exterior spaces. Also, within the house, subtle level changes

between rooms not only add interest to the interior layout but also reflect the site's topography.

In the private areas of the home, which are articulated through a long corridor, the occupants have access to the second garden, located at the highest point of the building, which is less formal and more recreational.

CG house 占地 1603m²，紧邻玛德雷山脉，陡峭的山坡地形给这个方案的设计带来了不少挑战，宽阔的蒙特雷斯市景观给建筑师提供了设计灵感。

沿街而过，两排巨大的橡树绵延到游泳池的露台，围合成了一个私密性很强的迎宾林荫道。根据天然坡地的地形，游泳池和花园被一个巨大的水泥墙围合起来，棕色、黑色火山石、IPE木、白粉刷盒钢铁的应用为房子设立了一个特有基调。

沿着宽阔的花岗楼梯而上，会看到一个棕榈树花园。正前方是一个3.6m高的黑橡木门，门且如此，整个990m²空间的大规模气势可见一斑。入口处是一个点缀着银叶的4.5m高的围墙，旁边有一个别具匠心的用椰子树根和钢铁、石头做成的雕刻。越墙而过就到了室内，首先映入眼帘的是客厅，客厅和卧室、餐厅一样有着大大的窗户，通过窗户可以看到美丽的花园。阳光透过窗户挥洒而下，室内和室外也因为同样的明亮而变得异乎寻常的融合。房间内部有着轻微的高低错落，漫步其中，惬意暗生，坡地的地形特征在屋子内部巧妙地表现出来。

此外，穿过一条长长的走廊，来到私密领域——第二花园，这个花园位于整个布局的最高点，摒弃了喧嚣，隔离了繁杂，一点点休闲，一点点闲适，使业主可以得到彻底的放松。

MÉXICO

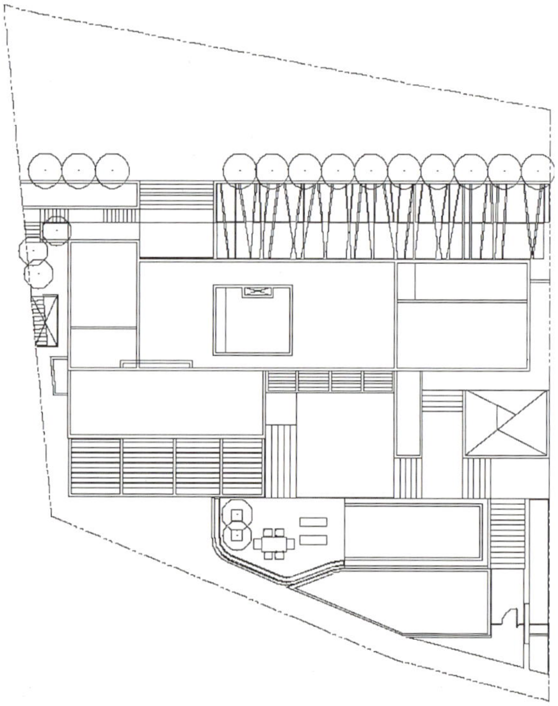

La Residence

拉居

Architect: John Friedman Alice Kimm Architects

建筑：约翰·弗瑞德曼·爱丽思·金姆建筑师事务所

Surrounded by stately residences in a variety of historical styles, this 5,300 square foot single family residence responds with a deceptively simple composition of volumes, voids, and screens. The balanced muscularity of the street facade presents a contemporary response to the formality and classicism of the adjacent structures.

At the rear of the house, the structure includes a kitchen/family room wing that extends into the private garden. Generously-sized terraces and large banks of sliding glass doors encourage year-round indoor-outdoor living.

The emphasis on simplicity continues inside the structure, which employs a free plan animated by a floating staircase located between the foyer and living room. Abundant use of natural light activates the interior spaces.

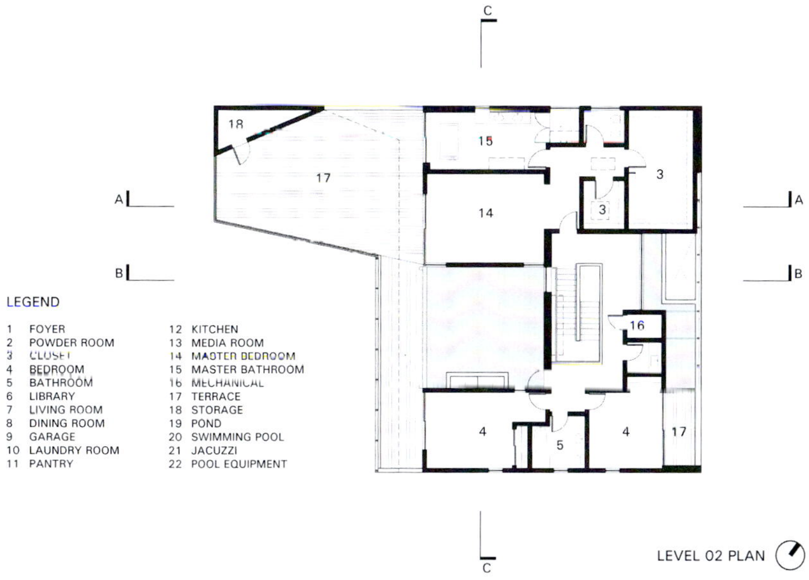

在各种庄严的历史风格住宅群之中，这座5,300平方英尺（492.37m²）的独户住宅让人迷惑，看似只是简单的空间、空白和屏风的结合。临街的外墙平稳而厚实，和相邻建筑物的拘谨和古典主义保持格调一致。

在住宅的后部，是一个厨房/亲子房，侧面延伸到私人花园。宽大的露台和大型滑动玻璃门台，使得户内户外生活永远舒适。

内部继续强调朴素的风格，通过门厅和客厅之间的移动楼梯，可以自由地进行空间布局。大量自然光使得内部空间栩栩如生。

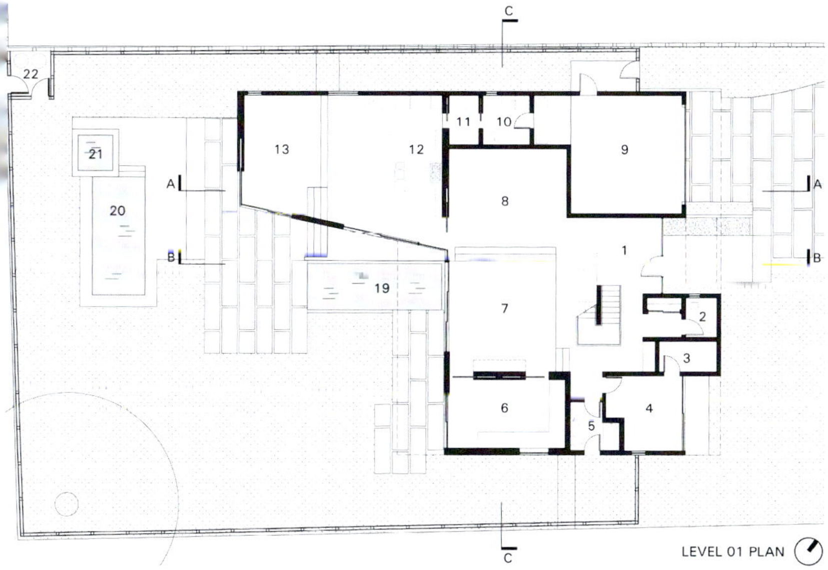

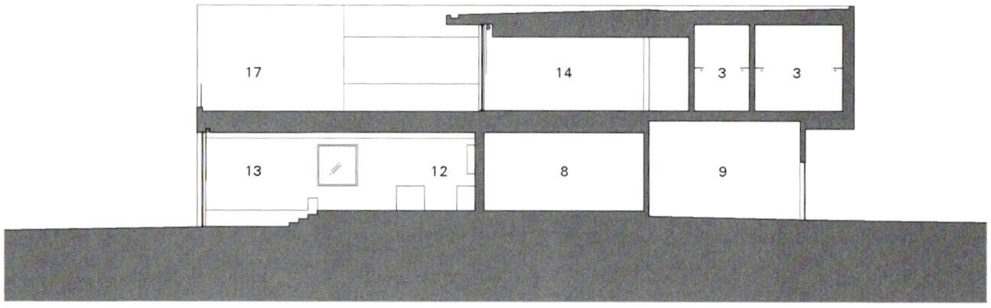

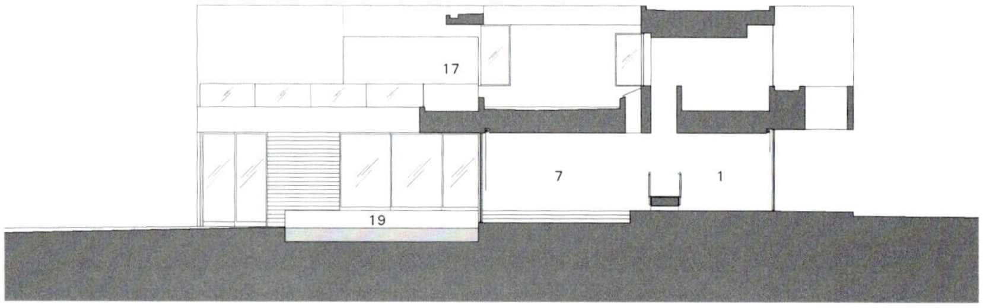

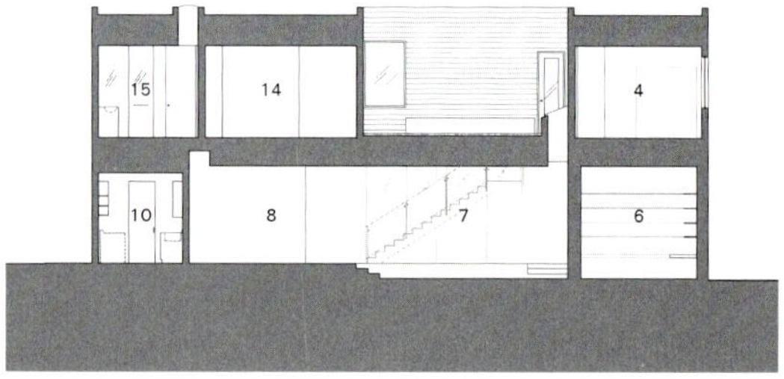

LEGEND

1 FOYER
2 POWDER ROOM
3 CLOSET
4 BEDROOM
5 BATHROOM
6 LIBRARY
7 LIVING ROOM
8 DINING ROOM
9 GARAGE
10 LAUNDRY ROOM
11 PANTRY
12 KITCHEN
13 MEDIA ROOM
14 MASTER BEDROOM
15 MASTER BATHROOM
16 MECHANICAL
17 TERRACE
18 STORAGE
19 POND
20 SWIMMING POOL
21 JACUZZI
22 POOL EQUIPMENT

Lockyer Residence

洛克耶住宅

Building area: 95 m^2　Designed by: Shaun Lockyer Architects with Arkhefield

建筑面积：95m^2　设计：肖恩·洛克耶与建筑师 Arkhefield

The "Lockyer House" is a small, contemporary extension onto a post war house in Bardon, a heavily treed and hilly fringe suburb in Brisbane. The design looks to address two primary architectural issues, the first is about context and language, the second is about finding the "essence" of what is needed from an accommodation resource perspective in an effort to create a engaging but practical and economical outcome.

The extension is essentially a "pod" added onto the end of a post war house which, while largely leaving the original house in tact, allows for the overall outcome to be reprogrammed. The architecture is unapologetic in its overt, contemporary nature that sits in stark contrast to the original. Notwithstanding this bold move, the original proportions and integrity of the cottage are left in tact to preserve the virtue of both.

The planning, layout and architecture of the space is about the family, the connection to the landscape and the engagement with the sky. The space was to address the cold winters and to allow for increased summer breezes, both of which were inadequate in the original cottage.

The extension was an experiment in achieving a qualitative outcome at a very low cost. The process afforded a relatively high degree of

craft and joy in what is essentially a very simple, easy to construct little box.

The design of this house is all about creation of a simple, cost effective "pod" that needed to accommodate the needs of a young family that had outgrown the original cottage. The other core function of the extension was to improve the connection to the landscape, engage with the sky and to provide a more climatically comfortable place to live, using the minimum resources possible to achieve this. The space is about joy.

"洛克耶住宅"是一个小型的，对布里斯班郊外巴尔东一处战后修建的住宅的现代扩建，绿树成荫，远山如黛。设计要解决两个基本的建筑问题：一个就是环境和风格，另一个就是找寻"本质"，也就是说要尽力满足住宿的需要，保证经济实用。

扩建其实就是添加在原住宅后面的一个"豆荚"（小建筑群），大部分都明智地保持了住宅的原样，因为考虑到要改变整体效果。建筑风格，蓄意明显，现代化的气息与原住宅对比鲜明。尽管这一举动大胆，但是屋舍的原结构和整体性都得以保留。

空间的规划、布局和结构都围绕家庭与风景以及天空的衔接。这一空间是为了解决冬天的寒冷，增加夏天通风，因为原屋舍在这两方面都不完善。

扩建是为了尝试以低成本实现高效果。本质上一个非常简单、朴素随便的小方体，却蕴含着高品格的艺术与喜庆。

这座住宅的设计纯粹是简单地创造，低成本高效果的"豆荚"是为了满足在原屋舍长大的孩子，他们刚成家后的住宿需要。扩展住宅的其他核心功能就是为了增强与风景以及天空的衔接，使得在小气候上更加适合居住，尽可能的节省资源。空间充满喜庆氛围。

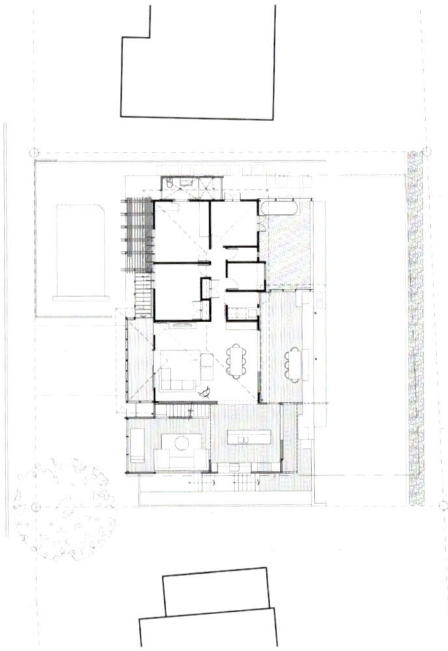

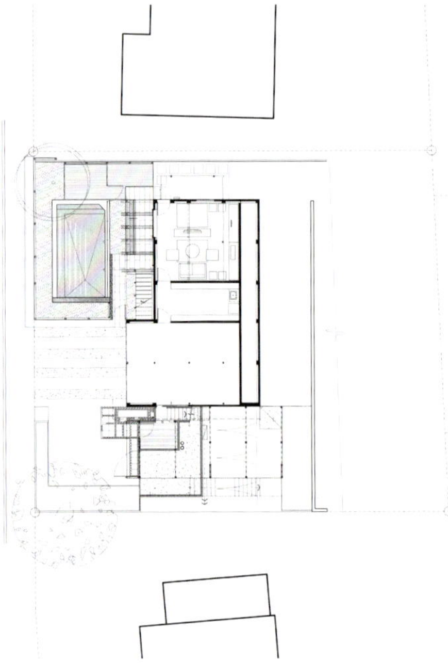

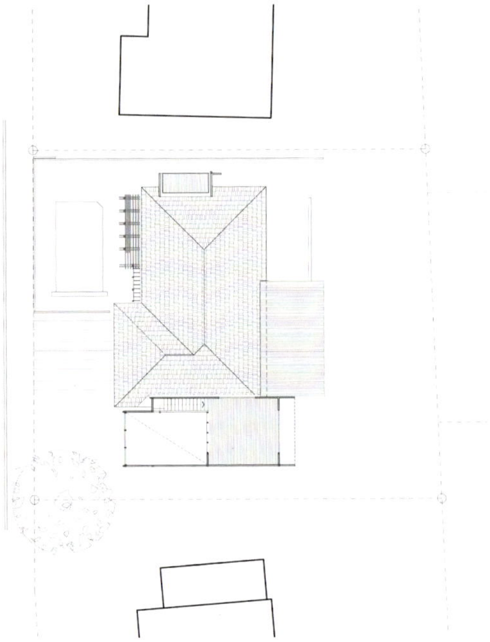

Villa Scianti, Modena (I)

摩德纳Scianti别墅（一）

Dimensions Plot area: 760 m²　Design architects: Davide Marazzi, Federico Pompignoli

总平面面积：760m²　建筑设计师：戴维德·马拉齐，费德里科·蓬皮尼奥利

The plan aims at enhancing the relation between interior and exterior space, securing large green areas to each floor. As a result, the house rises off the ground and open space flows continuously underneath it and ascends the house in succession becoming garden, portico, terrace and patio.

Out of the 665-sqm property, 625 sqm is restored to open air. In terms of composition, the scheme consists of three stereo-metric overlapping blocks slightly shifted from one another; the static paradox of the heavy volume supported by a transparent one gives the building a decidedly astonishing character.

Each block contains a different programmatic function of the house: the ground floor includes a garage, service rooms, a summer living area and a guestroom. The first floor, which is completely transparent by means of glass walls, is made of a large living room, the kitchen and a small office. The second floor, developed according to a patio scheme, encloses the sleeping area: here, for privacy reasons, the glazing is open only onto patios set up with plants and not to the environs.

The house will be entirely built of exposed concrete and glass; the concrete will have a grey shade in the lower volume, whilst the upper volume will be of a light terracotta shade.

该设计旨在加强内部和外部空间之间的联系，确保各楼层都有大块绿地面积。因此，房子高于地面，下面是连片的休憩用地，上面房子主要包括花园、门廊、阳台和庭院。

665m²的房产中，有625m²是露天的。在建筑组合方面，该计划由三个立体空间上彼此重叠而又稍稍错开的大楼组成，重型建筑和透明建筑之间的对比赋予大楼一种令人惊讶的特色。

房子的每一模块都包含一个不同计划功能：一楼包括车库、服务室、夏季生活区和客房。二楼全部是玻璃墙，透明光亮，包括大客厅、厨房和一个小办公室。三楼根据庭院计划设计而成，主要是卧室区：在这里，出于保护隐私的目的，只在栽满植物植物的庭院透光，周围地区不透光。

这所房子的外部全部由混凝土和玻璃建成，较低楼层混凝土为灰色，较高楼层的混凝土为淡红褐色。

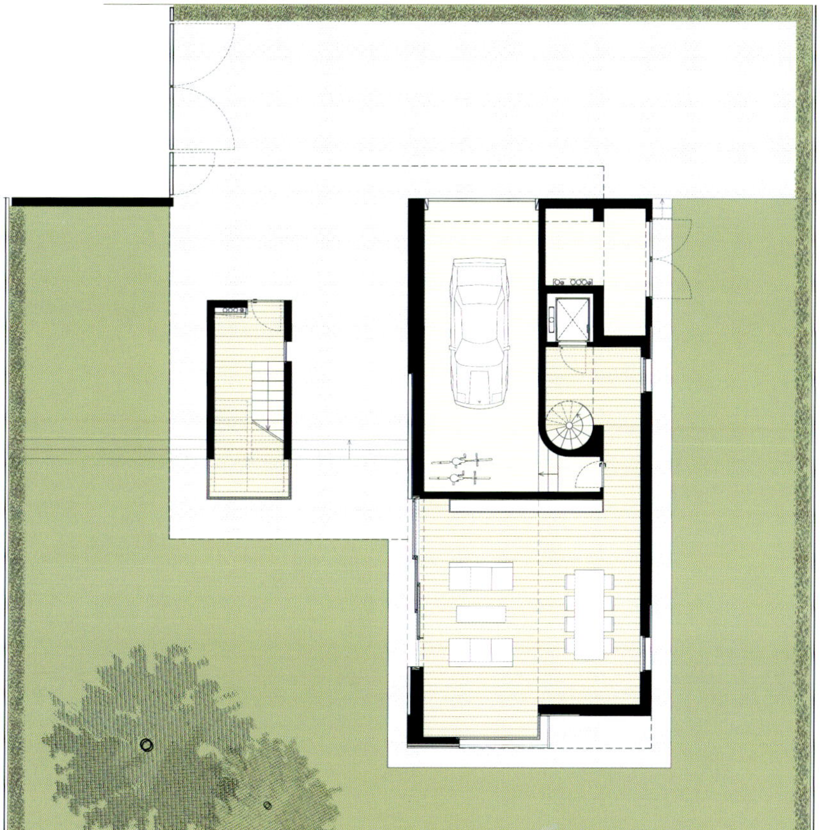

Meadowview

草原美景

Designed by: Platform 5　Designer: Peter Allen, Patrick Michell

设计公司：五号站台　设计师：彼得·艾伦，帕特里克·米歇尔

The site lies on the edge of a ribbon development village in rural Bedfordshire and is surrounded by mature trees, hedgerows and arable fields. A sweet chestnut clad box is cantilevered off a solid masonry and glass plinth, from across the fields, it looks like it is floating over the hedgerows. The different levels, walls and vegetation are used to vary the views. The site is divided up into a patchwork of different landscapes to create a transition between the domestic and agricultural environments. A private walled garden centred on a mature cherry tree is positioned on the more public side to the east. To the west, a raised terrace area hovers over a field of long grasses and meadow flowers.

Meadowview is a country house that is sensitive to its rural setting and incorporates many sustainable features. The route through the house gives a sequence of gradually more private spaces, both internal and external, and frames views to the rear garden, courtyard garden and to the agricultural landscape beyond. The house and garden are conceived of as an extension of the landscape rather than an object placed in it.

The house is surrounded by deciduous trees which give shade in summer and let sunlight through to the house in winter. The courtyard garden has a wild cherry tree in its centre surrounded by a gabion wall filled with rough limestone. This will store the air cooled by transpiration in the tree'

s canopy during the summer and act as a heat store in spring and autumn with the wall giving shelter from the wind allowing the garden to be used for more months of the year.

该住宅濒临贝德福德郡农村发展带，周围环绕着参天大树、灌木树篱和大片田地。坚固石屋和玻璃柱脚外是悬臂支撑的甜美栗子包箱，远远望去，仿佛漂浮在灌木树篱上。层次各异、墙壁多样、植被丰富，景色错落有致。景观各有特色，家庭和农业环境不断转换。私人围墙花园的中心是一颗大樱桃树，坐落在更开放的东部。西部，高高筑起的露台矗立在高草与百花之中。

草原美景是乡村别墅，易受农村环境感染，综合了许多可持续的特性。不管是内部还是外部，走进别墅，空间越来越私密，后花园、庭院花园的框景别具一格，外面是广阔的农业景观。别墅与花园是景观的延伸，而非其中实物。

别墅周围是落叶树木，夏日带来阴凉，冬日带来温暖阳光。用粗糙石灰石的石笼墙围起来的庭院花园中央有一棵野樱桃树。夏天，茂密的樱桃树进行蒸腾作用，花园空气凉爽；春天和秋天，石墙挡住了冷风，花园空气温暖，可用的时间多出了几个月。

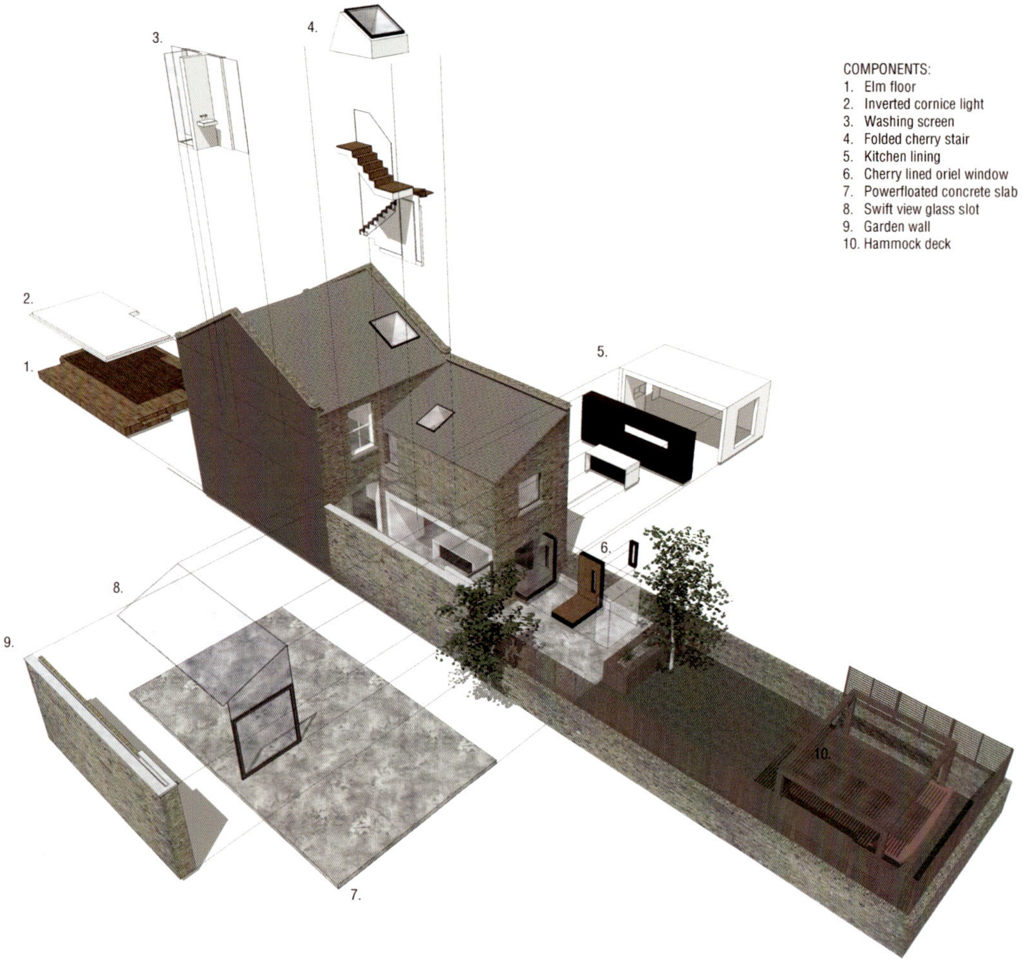

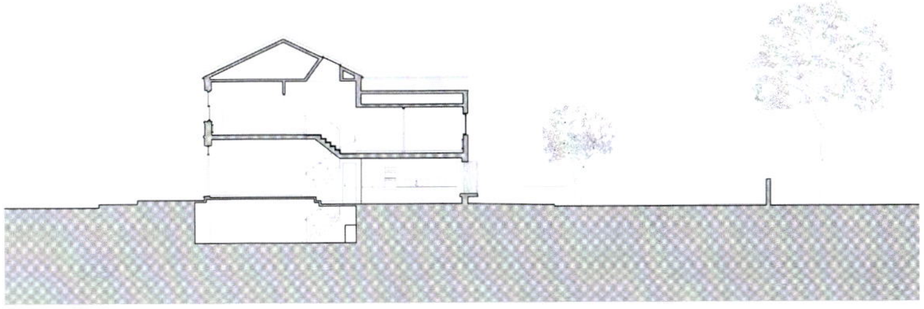

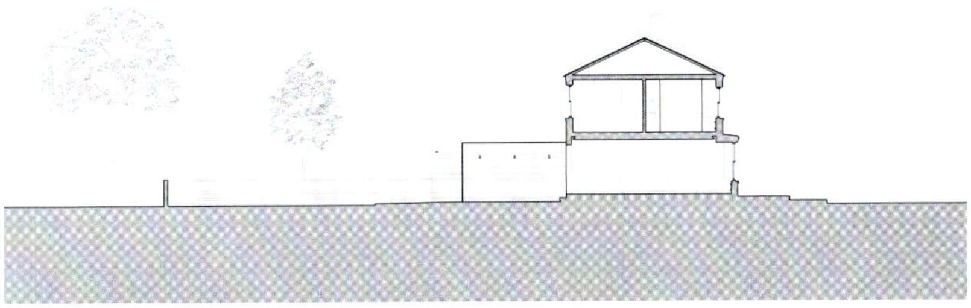

Versailles Residence

凡尔赛别墅

Architect: morrison seifert murphy　　Location: Dallas, Texas

建筑师：莫里森・塞弗特・墨菲　　位置：得克萨斯州达拉斯市

The owners purchased this lot because of its proximity to a beautiful public park across the street. To take advantage of this natural asset the architects placed the primary living spaces on the second floor overlooking the park and the ancillary spaces on the ground level.

The main living space runs the full width of the house and is subdivided by low elements that contain elements such as the fireplace, display and entertainment cabinets and bookshelves. The wall facing the park is full-height, faceted glass and affords panoramic views of the park and neighborhood. Clerestory windows on the other three walls create the effect of a floating ceiling plane from both inside and outside the space and heighten the sense of expansiveness within this great room.

The lower portion of the main facade is rendered in Texas limestone while the remainder of the house is finished in hand-troweled stucco. The roof, supported by steel pipe columns, overhangs the main facade to shade the glass wall from the south sun.

业主购买这一地点主要是因为它靠近街对面的美丽公园。为了利用这个天然资源，建筑师把基本生活空间放在二楼，以便俯瞰对面的公园，把辅助空间放在一层。

主要生活空间占据整个房子的宽度，又分为次元素，比如壁炉、展示柜、娱乐柜和书架等元素。面对公园的墙壁是全高的面玻璃做的，这样就可以把公园和附近的景色尽收眼底了。其他三面墙壁上的天窗创造了这样一种效果：空间内外的天花板像是漂浮的，增强了这个大房间的延伸感。

主立面的下部是得克萨斯石灰石砌成的，而房子其余的主立面是手抹的灰泥。由钢管柱支撑的屋顶悬于主立面之上，为玻璃墙遮挡南面的日光。

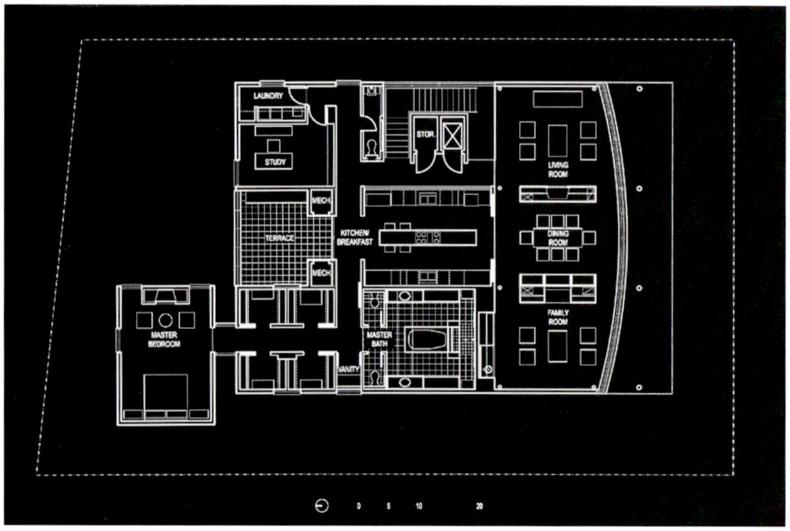

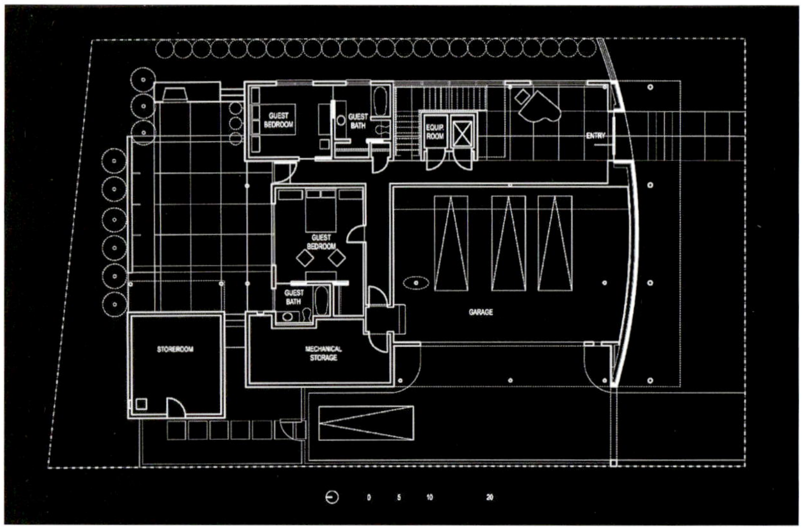

Pinball House

弹球屋

Floor area: 508 m^2 Location: Denmark

占地面积：508m^2 位置：丹麦

Idea: The villa echoes its surroundings closely and opens up towards the magnificent view over the lakes of Silkeborg. The facade resembles a lifted skirt that floods the terrace and primary living area in light. The broken line of the facade is continued into the central rooms of the house, where like the ball in a pinball machine it shoots around to draw dynamic balconies inside its square volume.

设计理念：该别墅与周围环境紧密相连，面朝景色蔚为壮观的施克堡湖区。建筑立面如同一条被掀起的裙子，将露台和主要生活区显露在光线中。截断的立面线条在别墅的主要空间中得到延续，如同弹球机中弹射出的弹球，在方形的建筑体上画出一个个动态的阳台。

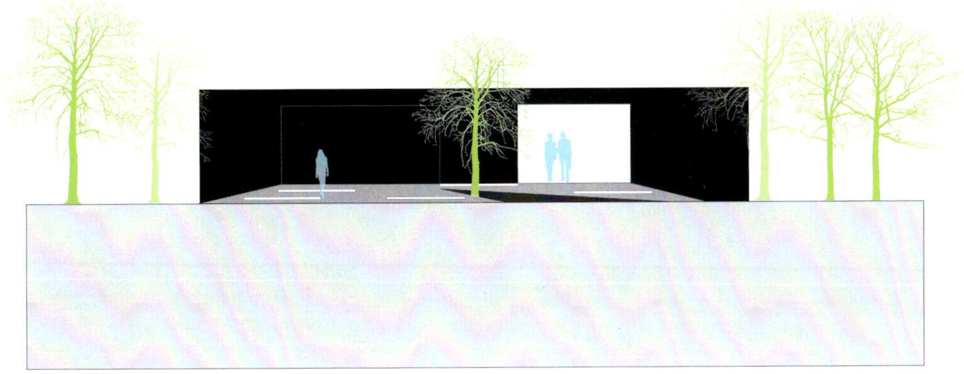

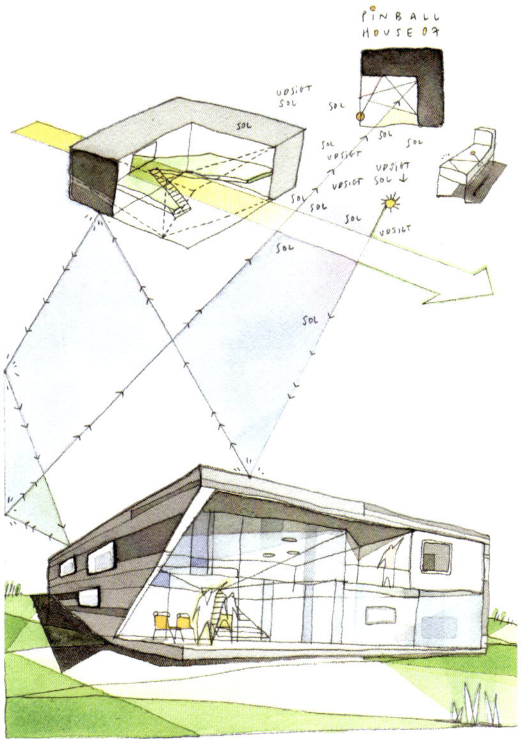
PINBALL
HOUSE 07
UDSIGT
SOL
SOL
UDSIGT
SOL

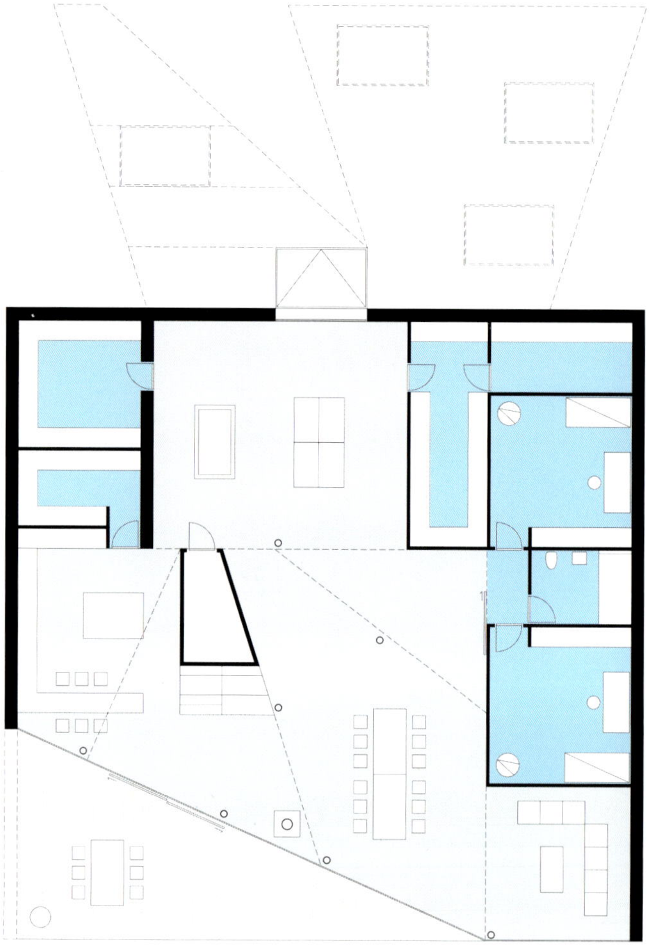

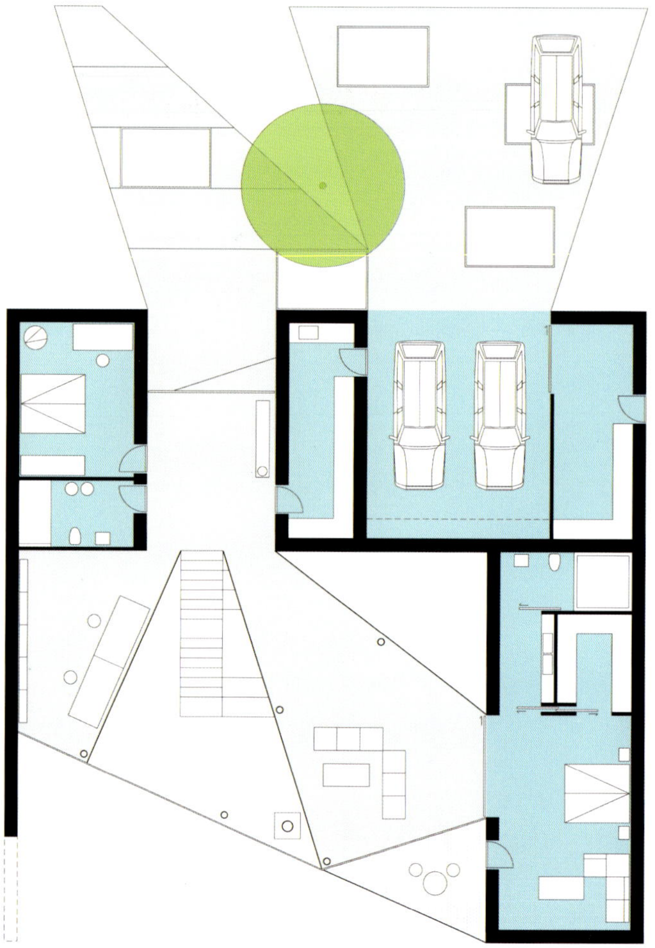

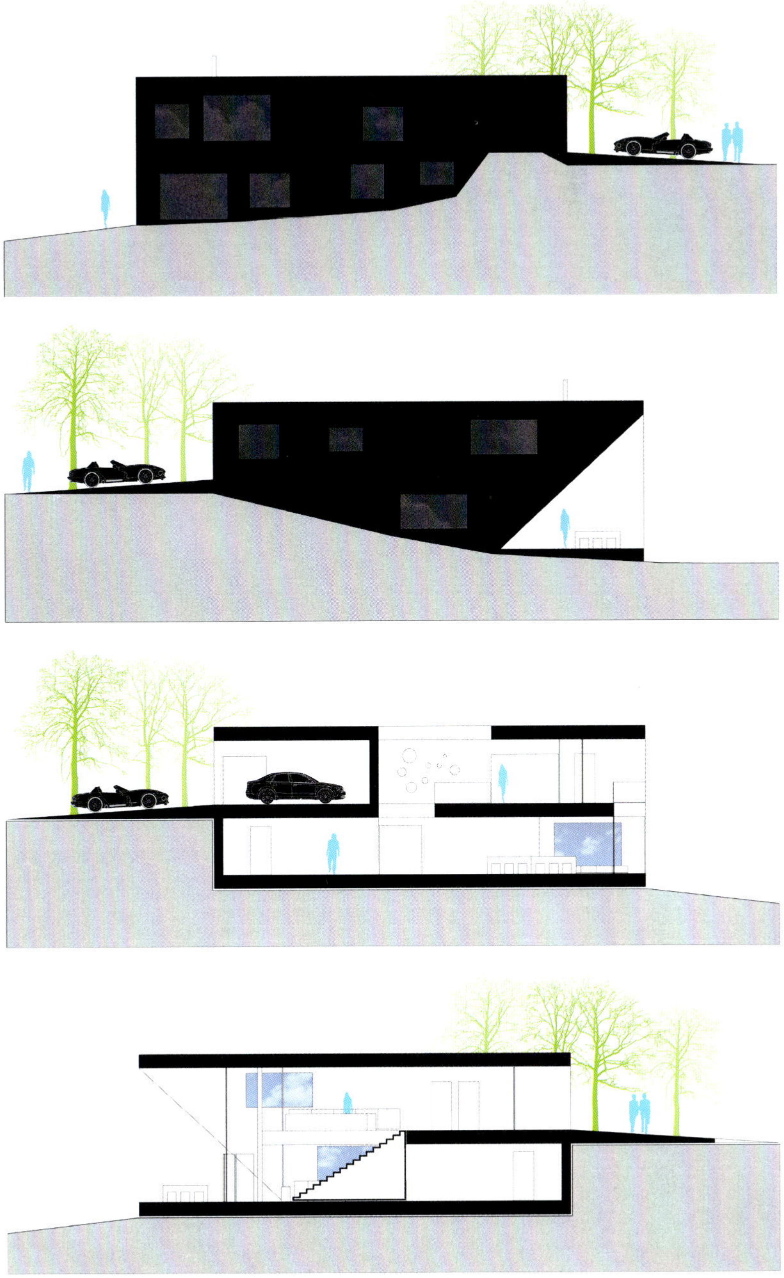

Cube House

魔方屋

Size: 220 m² Designed by: PLASMA studio, Lanerweg 18

面积：220m² 设计：等离子工作室, Lanerweg 18

The major constraints of this project, its steep site, compressed between existing buildings and very limited allowance of development volume have shaped the form of this house. It is inserted into the earth with 2 covered parking spaces to the front from where a small stair case leads up to the main living zones in the first floor and further to the bedrooms in the second floor.

Because of the limited available floor area the staircase and circulation had to be designed in a very space saving way – this lead to the continuous organization in the first floor: The single functions cooking, eating and living are positioned around the circulation core in order to give connectivity and privacy at the same time to the single activities - the staircase and built in furniture piece which is storage, oven and services cavity at the same time divides and connects as a short cut at the same time. In the second floor the single sleeping rooms are connected to each other in the shortest possible way.

Given prominent location of the site directed towards the South and the dolomites we opened the facade as much as possible in order to widen up the tights interiors – on both main floors ample balconies and terraces double the available floor area and offer great places to play for the kids and rest for the parents. In order to provide shelter from the views of the

1 living
2 cooking
3 eating
4 storage
5 office
6 stairs / corridor
7 balcony
8 terrace
9 bath
10 sleeping
11 entrance
12 cellar
13 parking

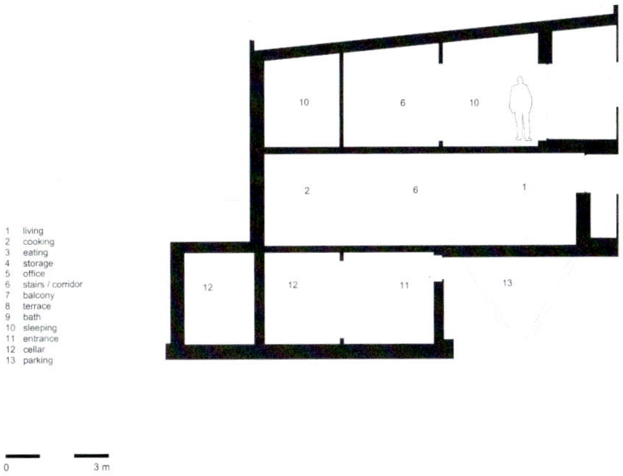

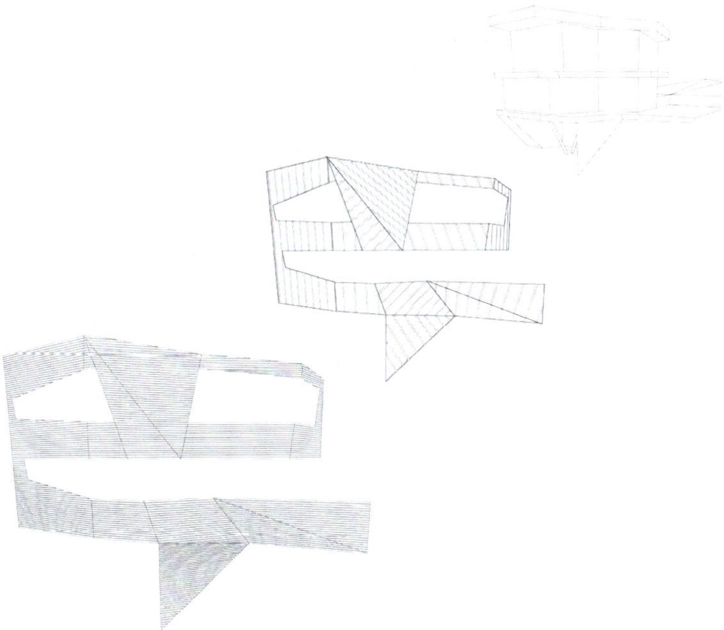

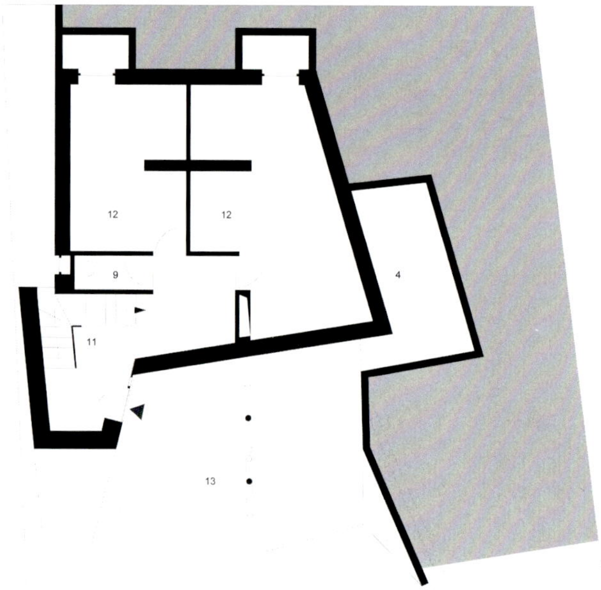

1 living
2 cooking
3 eating
4 storage
5 office
6 stairs / corridor
7 balcony
8 terrace
9 bath
10 sleeping
11 entrance
12 cellar
13 parking

0 3 m

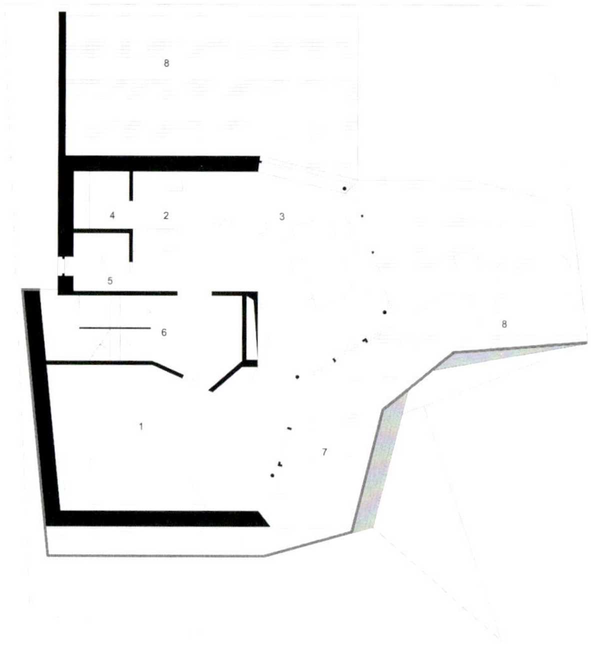

1 living
2 cooking
3 eating
4 storage
5 office
6 stairs / corridor
7 balcony
8 terrace
9 bath
10 sleeping
11 entrance
12 cellar
13 parking

0 3 m

passing- by road a layer of wooden sticks was wrapped around the big openings directed to the South – depending on the varying size of the openings they provide different degrees of shelter and intimacy.

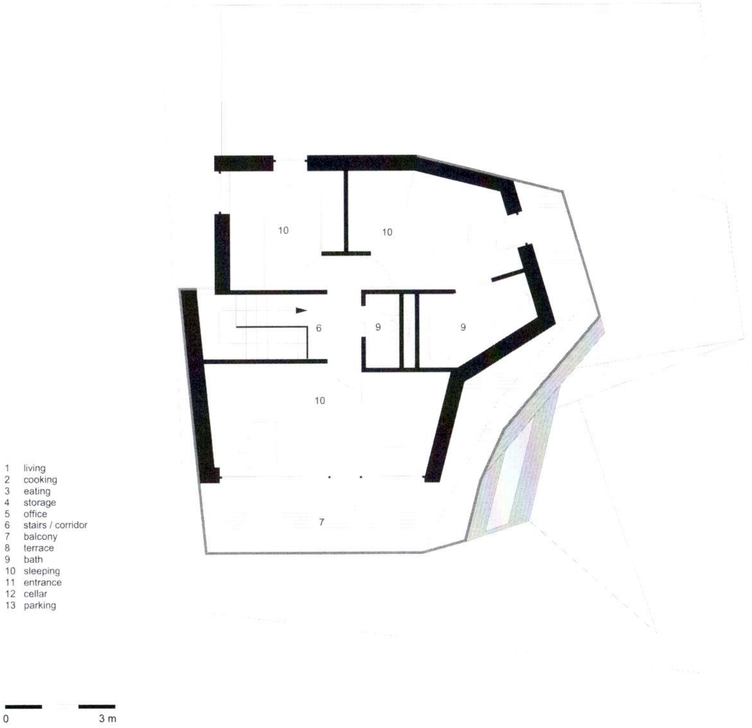

本项目的主要制约因素是陡峭的地形，建筑之间的空间比较小，不允许开发大面积的住宅，而且也影响住宅的造型。住宅前面有两个地下停车位，一个小楼梯从这里向上，通向一楼的主生活区，然后延伸到二楼的卧室。

因为可用建筑面积有限，所以楼梯和旋度的设计要能节省空间——这就要保持一楼的组织结构连贯。为了保持连贯和私密性的同时又能保证功能的正常使用，单一功能区，比如厨房、餐厅和客厅都被安排在旋度中心周围——楼梯和内置家具分别是储藏柜、炉灶和厨具柜，同时节省了空间。在二楼，单人卧室相互连接，也尽可能节省空间。

考虑到本案的正南方是白云石山脉，这一卓越的地理优势，所以我们把住宅正面尽可能地打开，以放大紧凑的内部——主楼层的两边，都有足够宽敞的阳台和露台，扩大楼层面积，为孩子玩耍和父母休息提供足够宽敞的场地。为了防止路上的行人偷窥，南边大门附近建了一圈木篱笆——根据门口的大小，确定不同程度的保护隐私的措施。

The Cats House

猫房

Design team: Eric Vreedenburgh, Coen Bouwmeester

设计团队：埃里克·威力登伯格，科恩·博文姆斯特

The house of family Cats is an example of a rooftop extension. Archipelontwerpers designed a lightweight steel construction on top of an existing modest office bungalow.

Because of the limited bearing capacities of the existing foundation, it was essential to do this project extremely lightweight.

The new volume contains an open flexible living and some terraces on different levels. The office lay-out of the first floor was transformed into the entrance hall and a bedroom area.

Archipelontwerpers used an easy industrial aluminium cladding system and a lightweight steel frame.

Except the kitchen and the stainless steel railing system, all the internal and external steel components are galvanized. The roof is rigged out with PV solar panels.

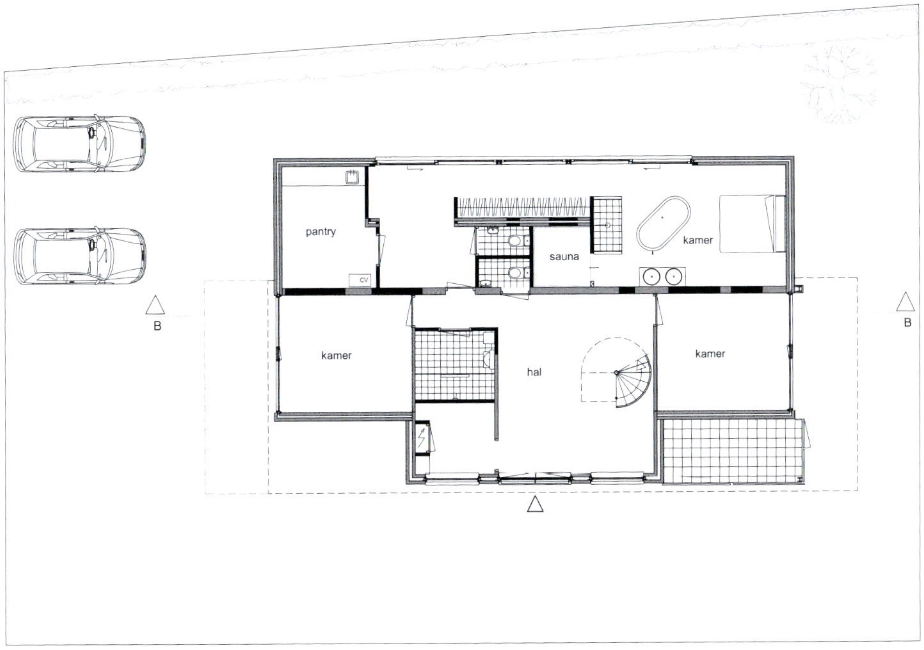

猫房是一个屋顶增建的例子。 Archipelontwerpers（设计公司名）在现有简单办公平房顶部设计了一种轻型钢结构。

现有基础的承载能力有限，必须使这个项目的重量非常小。

新建建筑包括一个开放灵活的居住空间和不同水平面的露台。一楼办公室布局改为入口大厅和卧室区。

Archipelontwerpers采用了一种简单的工业铝包层系统和轻钢框架。

除厨房和不锈钢栏杆系统外，所有的室内室外的钢构件都是镀锌的。屋顶安装了光伏太阳能电池板。

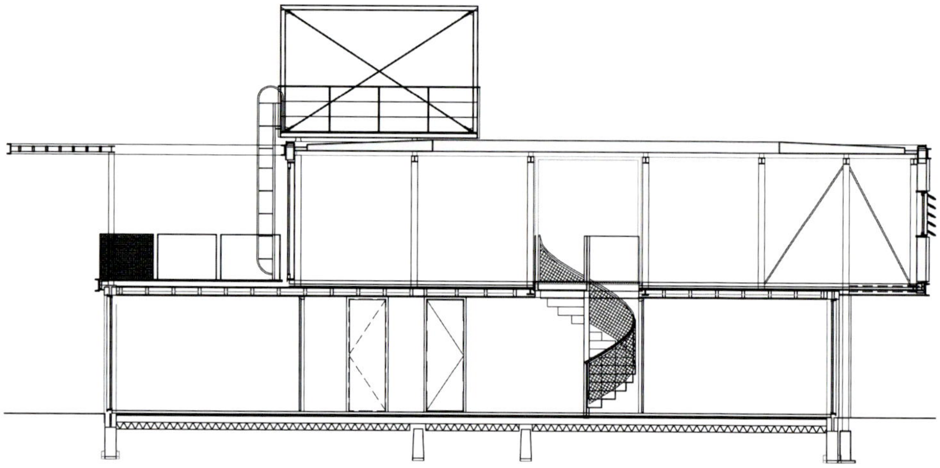

Torres House

托里斯住宅

Project Area: 5942 m^2 | Designed by: GLR arquitectos | Designer: Gilberto L. Rodriguez

建筑面积：5942m^2 | 设计公司：GLR建筑事务所 | 设计师：吉尔伯托 L. 罗德里格斯

Set high in the Sierra Madre mountain range, the Torres House seeks to establish an intimate contact with the natural environment in which it is located. Discreetly open towards the street, the house is accessed by crossing over a fountain from which a huge oak tree emerges, impressive both for its size and its beauty. Once on the porch, a grand mural-like door -constructed of marble, wood and steel- welcomes the visitor.

Inside, the entry hall merges into an area of contrasting materials consisting of Santo Tomas marble and volcanic stone walls. From this area there is a visual continuity with the rest of the house and with the garden that was left in its natural state as a small oak forest.

In the living room, conceived almost as a glass box, one can experience the view to the Sierra Madre peaks in the distance and the garden near by. Owing to its location within the mountain, where winters can get really cold, there is a fireplace for the family to gather around in the evenings.

From the dining room and the media room, one can access the roof terrace and enjoy the splendid views of the Sierra Madre as well as of the city of Monterrey. Here the musical background of the sounds of nature is in sharp contrast to the roar of the city below.

小屋位于马德雷山脉（Sierra Madre），周围自然景色怡人。房屋面向街道，入口处有一个喷泉，喷泉旁有一颗大橡树，枝叶繁茂，美丽异常，令人印象深刻。门廊处彷如壁画的大门，由大理石、石头和钢筋建造而成，欢迎每一位来访者。

入口处有一面色彩鲜艳的墙，它由圣托马斯（Santo Tomas）大理石和火山石建造而成。它给人以连续性的空间感，使门庭与房子的其他部分甚至室外的橡树园都融为一体。

客厅像一个玻璃盒子，从这里可以看到不远处马德雷山脉和周围花园的景色。冬天，这里的气候寒冷，因此，在房子中间有一个壁炉区供家人在夜晚聚在一块儿取暖。

走过餐厅和媒体室便可来到屋顶阳台，在这里也可以欣赏到马德雷山脉和蒙特雷城的无限风光。同时，在这里你也可以体会到如音乐般的自然同咆哮的城市形成的鲜明对比。

House at Shimogamo Yakocho

Yakocho下贺茂之屋

Floor area: 461.95 m² Location: Kyoto, Japan

总建筑面积：461.95m² 位置：日本京都

In order to enable the residents to feel a sense of spaciousness in the rather small 360 square meter corner lot tightly flanked on two sides by neighboring houses, the long applied concept of "Interface" was again used here.

Two kinds of screen of such Interface have been applied in this project. One is the circular, frosted glass screen enveloping and protecting the second story on the North-East. The other is the vertically louvered, smoked bamboo screen. While affording some level of privacy, this screen retains the silhouette of the houses next door and allows the residents of the new house to be aware of the neighbors.

The plan of the house is comprised of two bedrooms and related water closets on the ground level, a family room and a Japanese room flanking a sunken patio in an L-shape in the basement along with a study, a guestroom, and related sanitary rooms, a one-room style living, dining, and kitchen boasting over a 3-meter high ceiling on the second level, and a pergola-clad moon-gazing terrace up on the rooftop with a panorama of the hills of Kyoto all around.

The facade of the house is masked by a circular screen of frosted glass. Flanking this rather cold, high-tech expression are warm, natural, smoked bamboo louvers to strike a pleasant contrast between the industrialized material. The entrance approach and car parking area is lined with a paving of rusty granite, within which are planted bamboos and imbedded with accents of low-rise shrubbery and grass.

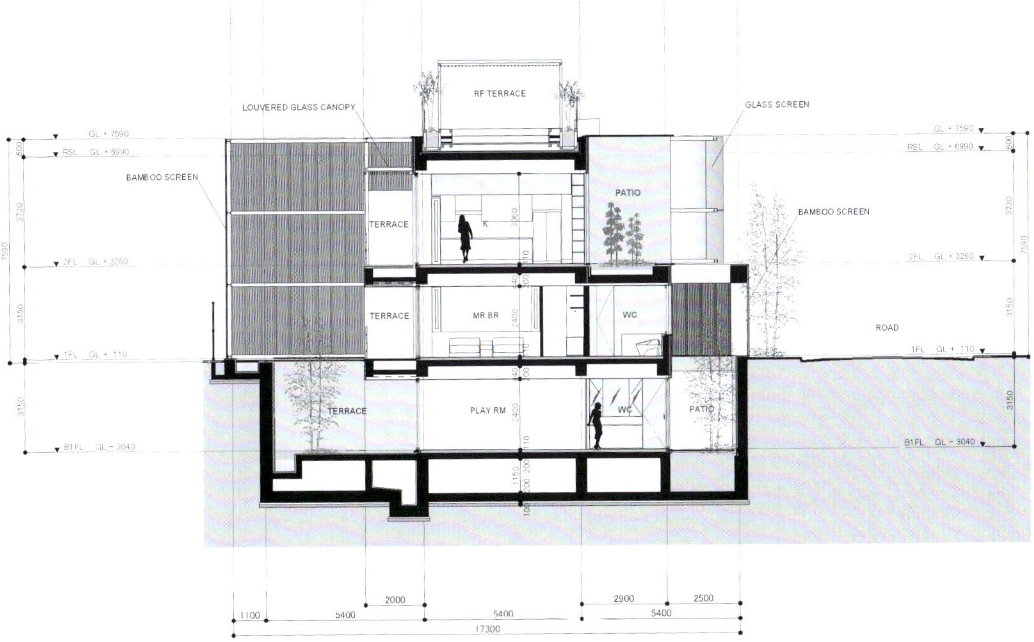
RF TERRACE
LOUVERED GLASS CANOPY
GLASS SCREEN
GL + 7590
RSL GL + 6990
BAMBOO SCREEN
TERRACE
K
PATIO
BAMBOO SCREEN
2FL GL + 3260
TERRACE
MR BR
WC
ROAD
1FL GL + 110
TERRACE
PLAY RM
WC
PATIO
B1FL GL - 3040
2000
2900
2500
1100
5400
5400
5400
17300

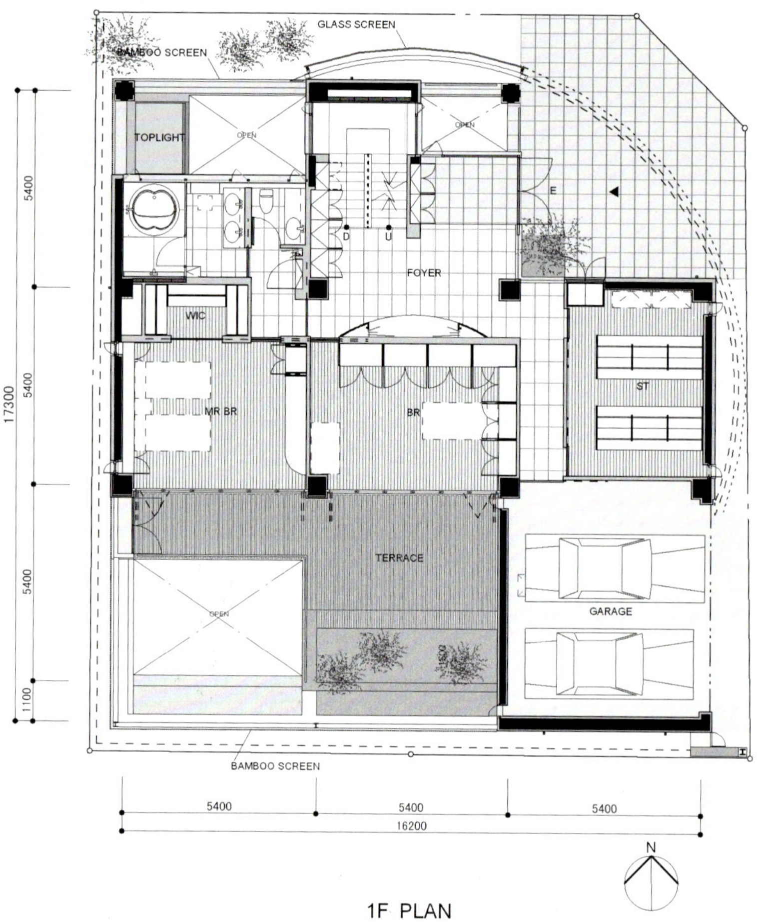

1F PLAN

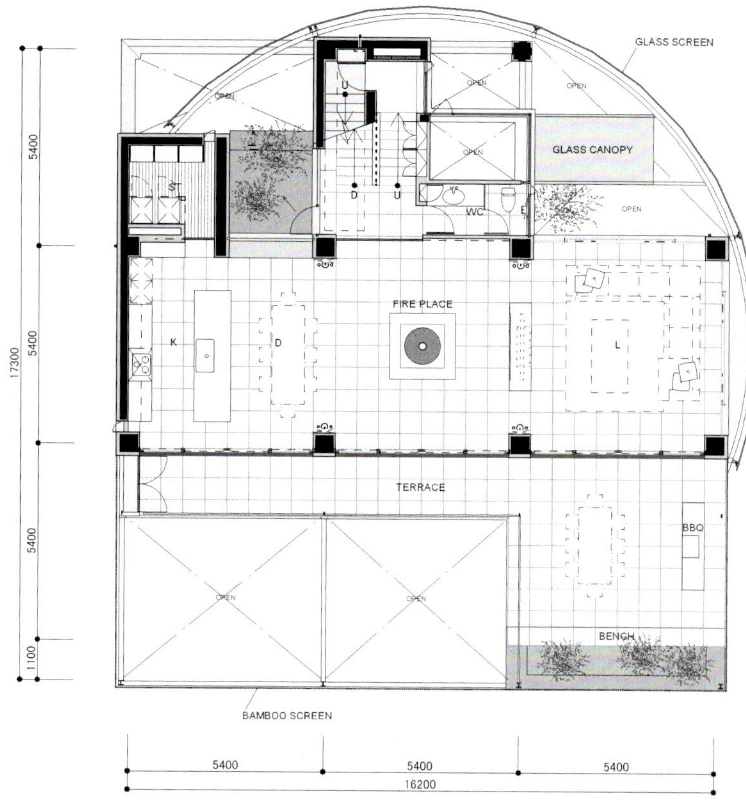

2F PLAN

该方案转角两侧都与邻居住宅紧密相邻，360m²的空间相当狭小，为了使居住者有宽敞明亮的感觉，这里再一次利用了一直以来沿用的“界面”的概念。

本项目应用了两种屏风界面。一个是圆形的，磨砂玻璃屏包覆和保护着二楼东北部。另一个是垂直的百叶式熏制竹帘。虽然在一定程度上，透露了隐私，但是透过竹帘，可以看到隔壁住宅的轮廓，让乔迁新居的住户认识邻居们。

住宅的一楼设计有两个卧室，都带相应的洗手间，与“L”型的凹陷天井侧面相接的一间亲子房和一间日式房，连同书房、客房以及相应的卫生间，单居室风格的客厅、餐厅以及厨房都在二楼，值得称赞的是，天花板的高度超过3m，从爬满藤蔓的赏月廊沿台阶走到屋顶，京都周围的丘陵一览无余。

住宅的外墙用圆形磨砂玻璃屏风加以掩饰。为了打破这种冷静感，巧妙地采用温暖的、自然的熏制竹帘，来冲击工业化材料，带来舒适美观的感受。入口处和停车区，两旁都用锈色花岗岩铺砌，里面种有竹子，矮灌木和草坪点缀其中。

TERRACE
STUDY
PLAY ROOM
GUEST ROOM
STORAGE
UTILITY
TERRACE
JAPANESE ROOM
PIT

Villa G

G公馆

Area: 368 m² Designer: Todd Saunders Interior designed by: Saunders Architecture

面积：368m² 设计师：Todd Saunders 室内设计：Saunders Architecture

The stair is one solid piece of 1cm thick steel, galvanised with white sand corn making it slipping-resistant. The washroom and storage spaces are integrated into thick walls that are at least 60-70cm deep. The kitchen bench is 8 meters long and has plenty of drawers for kitchen equipment and even other things that need to be stored away. None of the electrical outlets are visible and all technology controlled by a main control panel in the kitchen.

楼梯由1cm厚的钢板组成，上面镀上白砂粒，能起到防滑的作用。洗手间和储藏间都设计在厚厚的墙壁里，至少有0.6~0.7m深。厨房的烹饪台有8m长，上面还设计了很多抽屉，用来摆放和储藏各种厨房用品。厨房里看不到任何电路接口，同时通过操纵主控制板就能控制厨房内所有的设施。

1. ETG _1:200

1.floor plan:

1. Entrance
2. Hall
3. Living Room
4. Bedroom 1
5. Walk-in Closet
6. Bedroom 2
7. Bedroom 3
8. Bathroom
9. Storage/ Utility Room
10. Washroom
11. Garage

TOTAL AREA: 208 m²

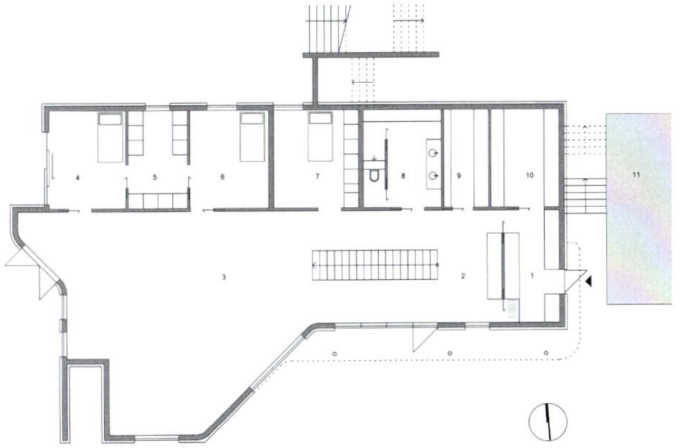

2. ETG _1:200

1.floor plan:

1. Terrasse
2. Living Room
3. Kitchen
4. Bathroom
5. Bedroom
6. Walk-in Closet
7. Bedroom
8. Storage
9. Opening under skylight

TOTAL AREA: 160 m²

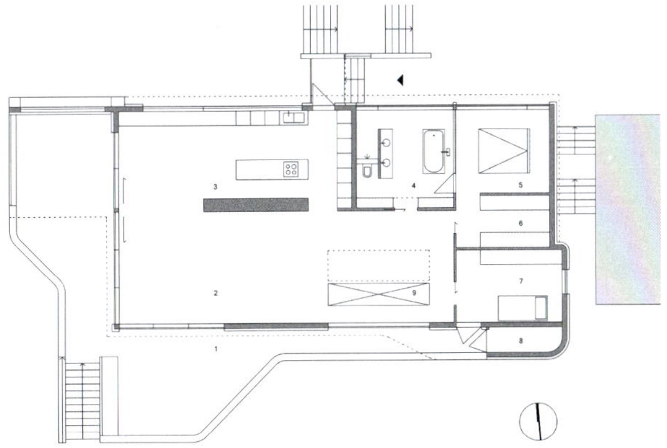

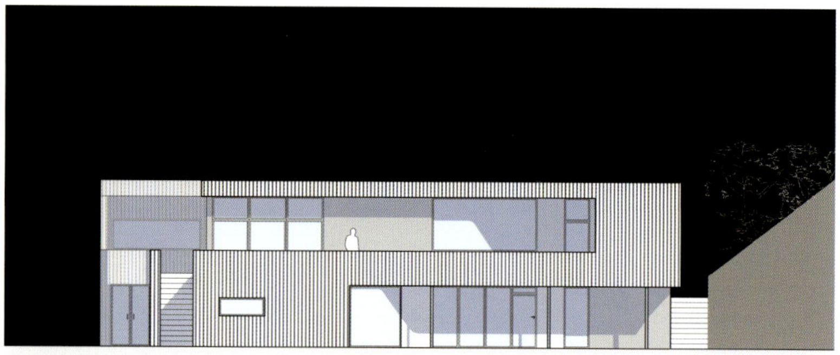

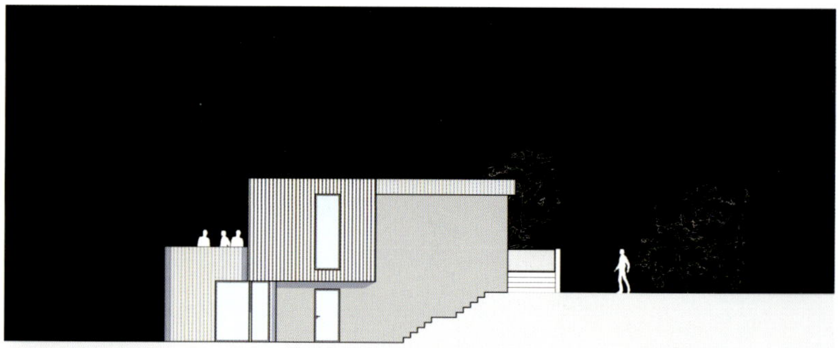

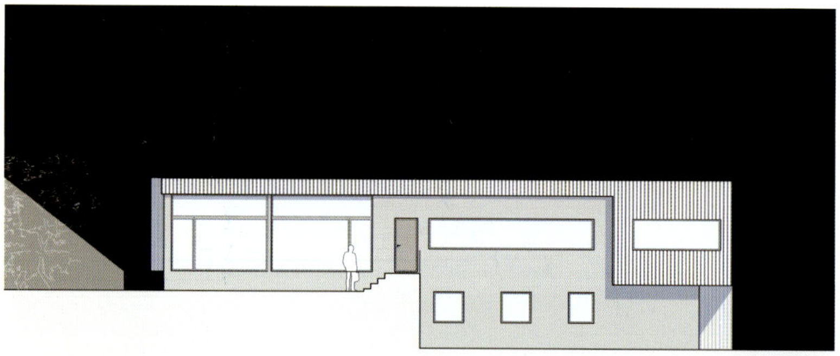

Villa Storingavika

Storingavika别墅

Area: 304 m² Designer: Todd Saunders Interior designed by: Saunders Architecture

面积：304m² 设计师：Todd Saunders 室内设计：Saunders Architecture

Overlooking breathtaking fjords and a stretch of Norway's west coast archipelago, Villa Storingavika is a robust yet refined vessel from which to appreciate the delicate coastline and sometimes rugged climate. It is a pale timber volume enrobed in a crisp, 'pleated' dark timber exterior. On the upper level, there are a generous open-plan living room and a master suite, while a lower level would be used by the owners' children and include a small self-contained apartment. The design also accords with the owners requirements for details, such as no trim around doors and windows and no baseboards where floor meets walls.

Storingavika别墅坐落于一个最具吸引力的位置上，能够俯瞰到南部的峡湾和挪威西海岸群岛，感受到偶尔严酷的气候。它看起来就像是一堆浅色的木料"穿"着深色的木质外套。楼上设计有一个开放的客厅和一套主卧室，而楼下则是一套独立的公寓，供户主的子女使用。设计中也遵循了户主一些细节上的要求，比如屋子的门和窗都不带任何装饰，而地板和墙相接之处不要基线板等。

ARKITEKTUR

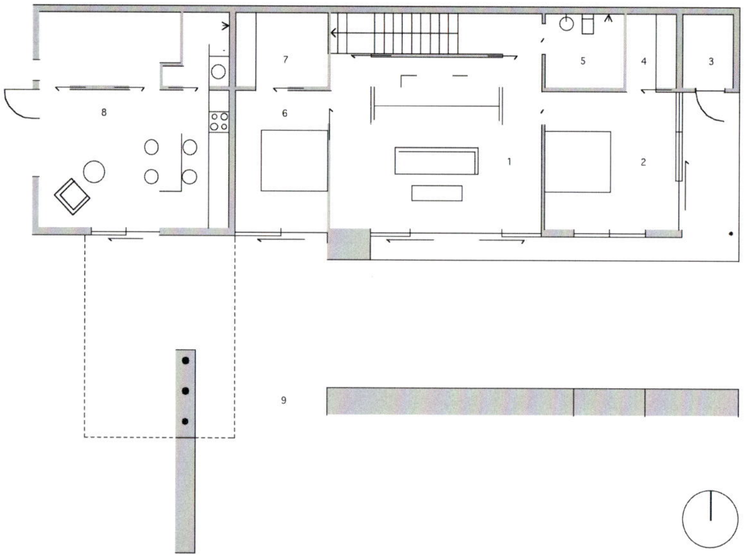

1.floor plan:

1. Livingroom with studio space
2. Bedroom
3. Storage
4. Walk-in Closet
5. Bathroom
6. Bedroom
7. Walk-in Closet
8. Apartment
9. Outdoor Furniture

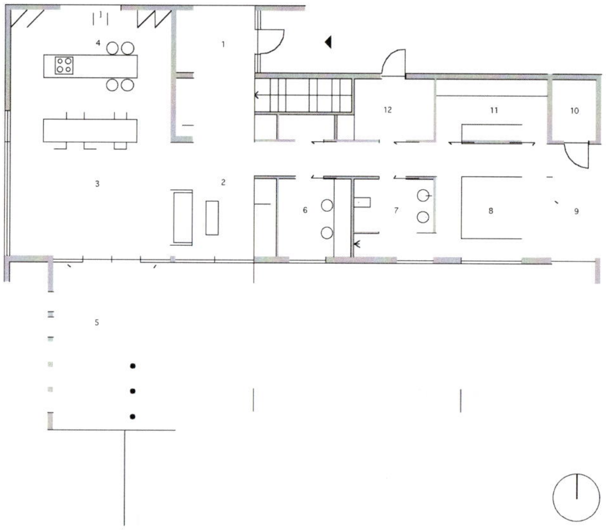

2.floor plan:

1. Entrance
2. Fireplace
3. Livingroom
4. Kitchen
5. Terrace
6. Office/ Guestroom
7. Bathroom
8. Bedroom
9. Terrace
10. Storage
11. Walk-in Closet
12. Utility Room

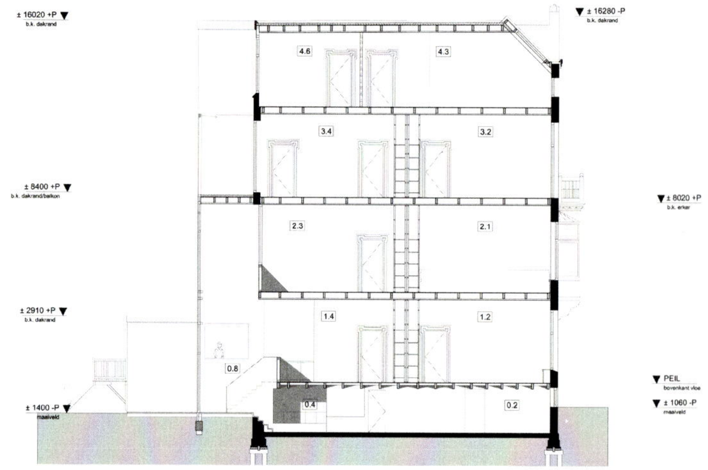

Mansion Alteration Amsterdam

阿姆斯特丹的改建大厦

Unfortunately the stately houses in the district Oud-Zuid in Amsterdam are more and more transformed into office spaces. For this reason it's very special that this house, at number 23 was transformed back from office to its original living function.

The mansion is carefully renovated and a series of spatial interferences are made. A unique family house of 500m2 arose. Because communication and sight are such significant elements to make a family house function, the client requested to "connect" different living functions but are not in each other's sight directly. This resulted in the most important adjustment in this house; a void over 3 stories.

The original conservatory and the back façade is broken down to create more space for the void over three floors. In this way the basement, the ground floor and first floor are connect and day light can enter deep into the mansion.

The new façade on this back side is made of large glass plates on which a pattern is sandblasted that shows similarity with the original fenestration; a modern façade with a clear reference to the history.

Over the different layers a new sculptural element is realized, that accommodates different functions: a home cinema and music

rehearsal, a lounge corner, staircases, cupboards and a fireplace. Behind the classical front façade of the mansion lies a very special spatial experience...

阿姆斯特丹老南城区庄严的住宅，几经改建，都变成了办公场所。正因为如此，这座住宅是第23栋从办公楼改建恢复原住宅功能的，从而显得非常特别。

经过精心装修和一系列的空间改造，一座500m²的独特住宅诞生了。因为沟通和景象是住宅功能非常重要的元素，所以客户要求各种居住功能互相“衔接”，但又互不影响。这就要对住宅进行最重要的调整；一个三层楼高的空间。

毁掉原来的暖房和后墙，给三层楼高的空间营造出更多的空间感。这样，地下室、一楼和二楼就相通了，日光可以照到底层。

后面的新墙采用大型的磨砂玻璃板，和原来的门窗布局相似；现代化的外墙，是历史清晰的参照。

各个层次，都采用了新型雕刻元素，体现不同的功能：家庭影院和练歌房、休息室、楼梯、碗橱和壁炉。古典式住宅门后，是非常特别的空间体验。

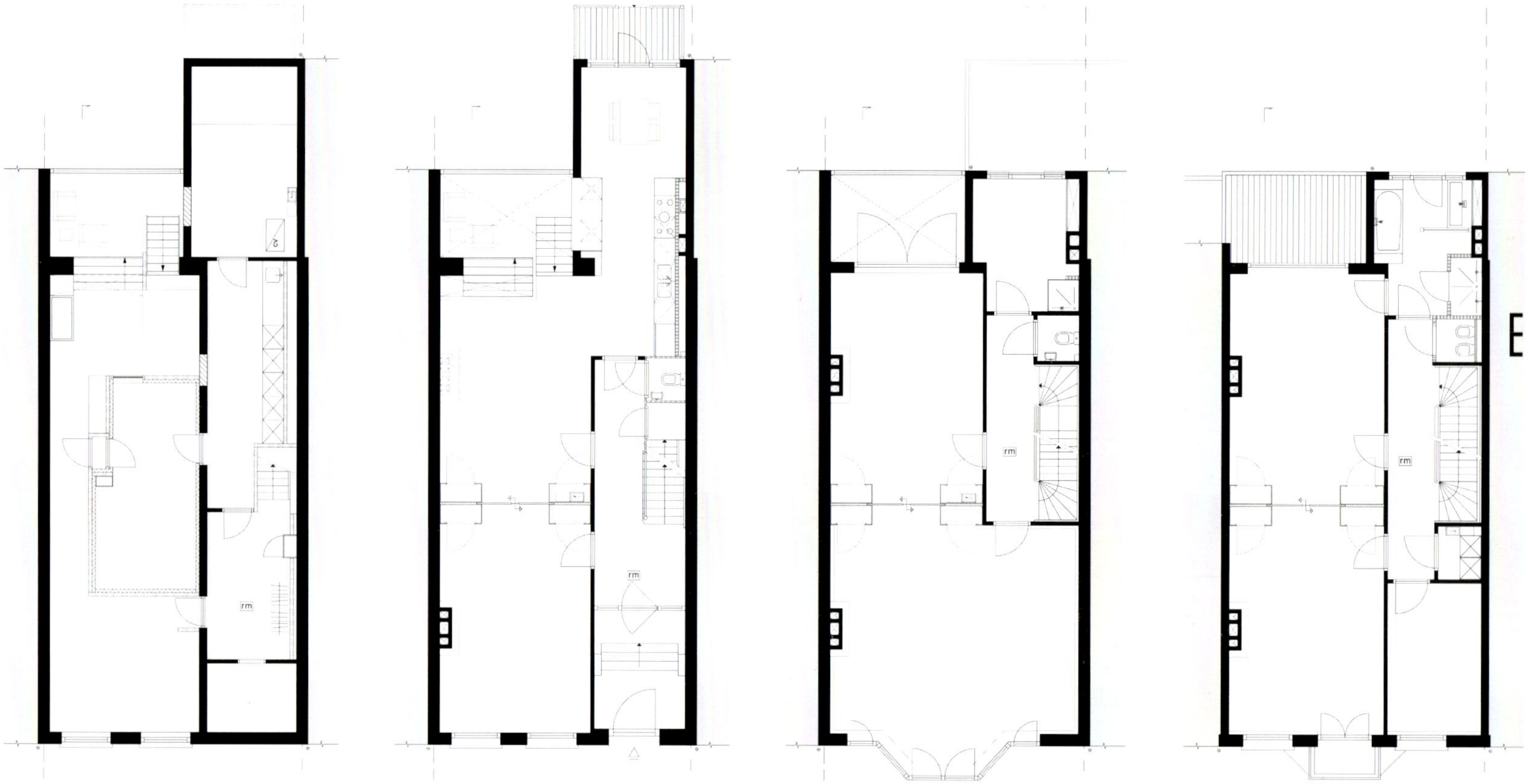
rm
rm
rm
rm

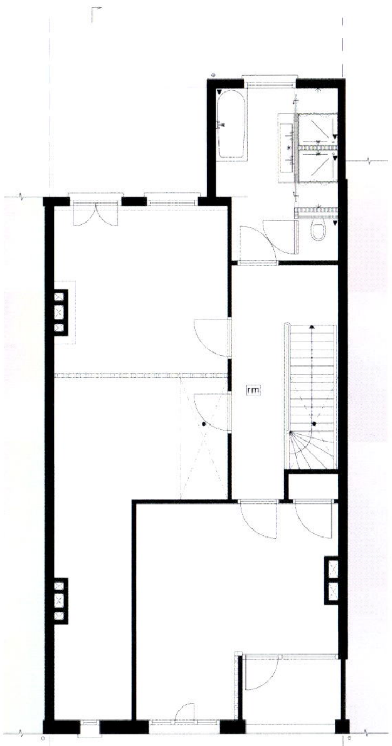

Villa Vriezenveen

Vriezenveen别墅

Location: Eastern Netherlands

位置：荷兰东部

The dark silhouettes of the historic farmhouses provided the most important source of inspiration for the statuesque house in Vriezenveen. The linear layout of the plots also influenced the design of the house. As in the big farmhouses of old, family life centres on the dining table in the large living room/kitchen.

The open landscape either side of an ancient country lane between Vriezenveen and Almelo forms the backdrop to this house. The contours of the sections, which are linked one behind the other, come together to form a unified, statuesque whole. The house is lengthened by the formation of the upper floor in two separate elements that are offset towards the front and rear ends. This produces a silhouette of apparently separate sections that stands out in the surrounding area. In addition, this formation leads to the creation of the requested covered terrace at the rear and a carport at the front.

This house follows the linear structure of the landscape. The black zinc of the outer walls is applied in long, horizontal strips with folded seams. Due to the sheen and somewhat uneven finish of the material, the house reacts in a subtle way to the colours of the Dutch sky. Thanks to the dark, subdued colour, the house blends in with the landscape more than the big farmhouses with their orange roofs.

The rooms of the house, with their large windows, are oriented towards the surrounding area. Each room has a strong focus on one of the points of the compass. Only the kitchen has a multiple orientation and functions as the central hub of the house. All the rooms are accessible from this central space, some of them via open staircases. The main load bearing structure consists of three steel portals supporting two laminated beams, each 1 metre high and 27 metres long! In combination with walls of a wooden skeleton structure, this allowed overhangs in multiple directions to be created.

农庄历史悠久，黑暗的轮廓是Vriezenveen别墅灵感最重要的来源。小块土地的线性布局也影响了房子的设计。在宽大的农庄里，人们在巨大的客厅或厨房里围着饭桌享受着古老传统的家庭生活。

Vriezenveen和阿尔默之间的古老乡村道路两旁的开阔景观给该房子提供了背景。各个部分的轮廓前后连接在一起，形成一个统一、高大的整体。两个独立元素的上层楼拉长了房子的长度，和前后两端相互呼应。明显独立部分的轮廓在周围地区显现出来。另外，这种设计创造了后面的封闭阳台和前面的车库。

这所房子遵循该地区景观的线性结构。外墙应用了水平带折缝的黑锌长线。这种光泽和有点粗糙的材料的房子使荷兰天空的颜色很微妙。多亏暗淡柔和的色彩，房子比有橙色屋顶的大农庄更好地与景观融为一体。

房子里带大窗户的房间都面向周边地区。每个房间都明显聚焦于某一个方向。只有厨房方位多端，并且是房子的中心枢纽。从这个中心区域可以进入所有房间，其中一些带有露天楼梯。主要承重结构为3根钢柱支撑的两根条状横梁，每根横梁高1m，长27m！组合使用木骨架结构墙体，创建了多个方向的交错重叠。

Villa Soest

苏斯特别墅

Program: Detached villa in a woods

方案：树林里的独立式别墅

Location: Soest, the Netherlands

位置：荷兰苏斯特

On a wide constructing site in the woods of Soest a new villa will be build. An important theme while designing the house is spatial connection between the three floors. From basement to the second floor a void has been planned.Standing at the front entrance a view over the void, directly into the garden, is provided. Functions as the toilet, wardrobe and closets are connected to this void. The parapet around the void continues smoothly into these functional spaces and make sure a sculptural link between the three floors is created.

在苏斯特森林里宽广的地面上将建立一座新的别墅。设计这个别墅的一个要点是三层楼之间的空间联系。一楼到二楼之间有一个空隙，站在前门，越过空隙可以直接观赏到花园。洗手间、衣柜和壁橱等功能连接到这一空隙区域。空隙周围的栏杆流畅地连接到这些功能空间，创造了三层楼之间的雕塑连接。

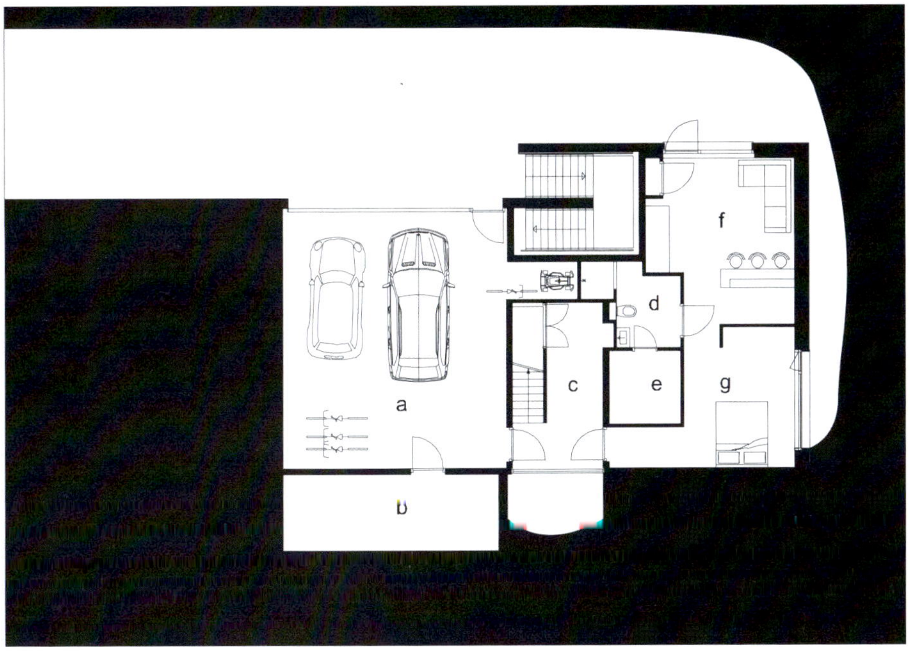

a garage
b storage
c hall
d bathroom
e sauna
f guests
g bedroom
h livingroom
i studyroom
j toilet
k dogkennel
l kitchen/ diningroom
m walk-in-closet

a garage
b storage
c hall
d bathroom
e sauna
f guests
g bedroom
h livingroom
i studyroom
j toilet
k dogkennel
l kitchen/ diningroom
m walk-in-closet

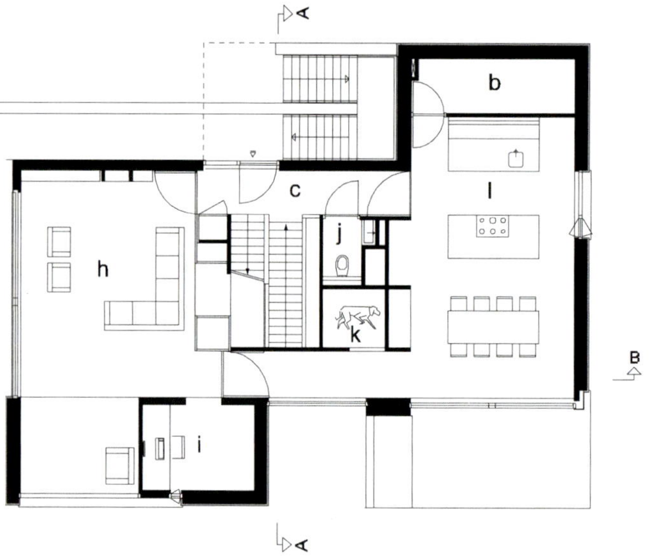

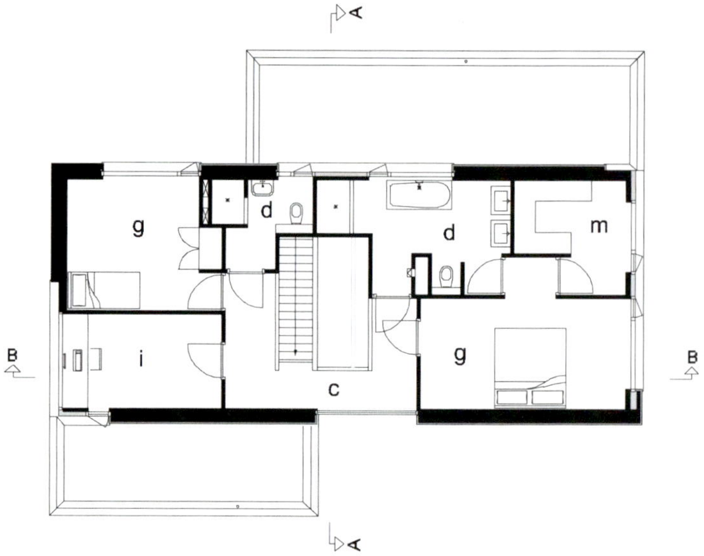

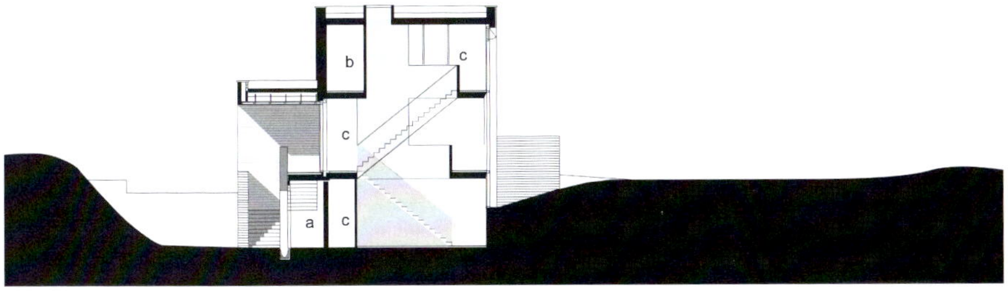

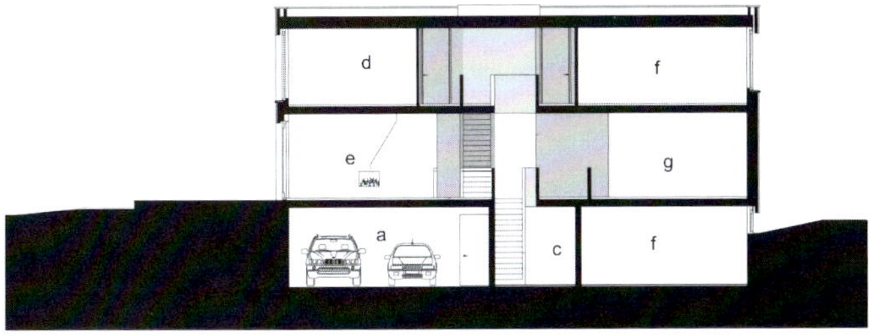

a garage
b bathroom
c hall
d studyroom
e livingroom
f bedroom
g kitchen/ dining room

Zeidler Residence

蔡德勒别墅

Architect: Ehrlich Architects　Interior designer: Arkhefield

建筑设计：欧利希建筑公司　室内设计：Arkhefield

Zeidler Residence is located in a quiet community, facing the sea, with beautiful environment. The residence consists of two independent low buildings which form a courtyard. The main building is designed with three floors according to different functions. The first floor is used as meeting room and recreational space. There are two spacious rooms: one is set with chairs, and the other leads to the kitchen and dining room. In the dining room, antique built-in wine rack to collect thousands of brand wines. On the second floor, there are mater bedroom with French window and balcony to enjoy sea view and daylight. The third floor is used for enjoying landscapes, recreation and storing equipments. A door is set at one side of the building, thus people can go upstairs directly.

The attached building is two-storey high with many guestrooms. It connects with the main building through the French window. The residence not only provides convenience to people but also creates a kind of high-quality lifestyle.

451

451

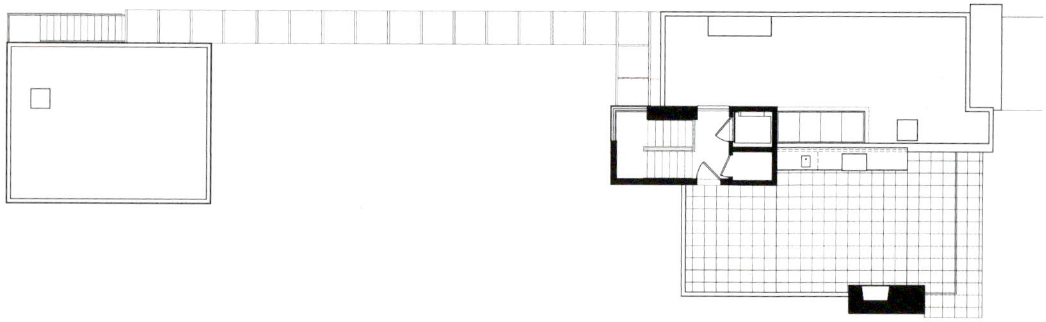

住所座落在安静的小区里，朝向海景，环境优美。由两栋独立的低层建筑构成，围合成院落。主楼按功能设计为三层，一层为会客、休闲场所，设置了两个宽敞的空间，一个可摆设配套的座椅，另一个通向厨房及餐饮区。餐饮区特别设计了品酒区，古色古香的墙壁酒架可收藏上千种名酒。二层为主人卧室。落地玻璃前设计有外飘的阳台，海景、阳光尽在眼前。三层为露天观景、休闲场所，可放置休闲设备。在主楼侧面另设计一个门，主客人不需经过会客大厅就可通向楼梯。

附楼为两层，设计了多间客房。通过落地玻璃窗的开合，主楼与附楼联系起来。住所在追求高品质生活的同时，也营造了便捷的生活方式。

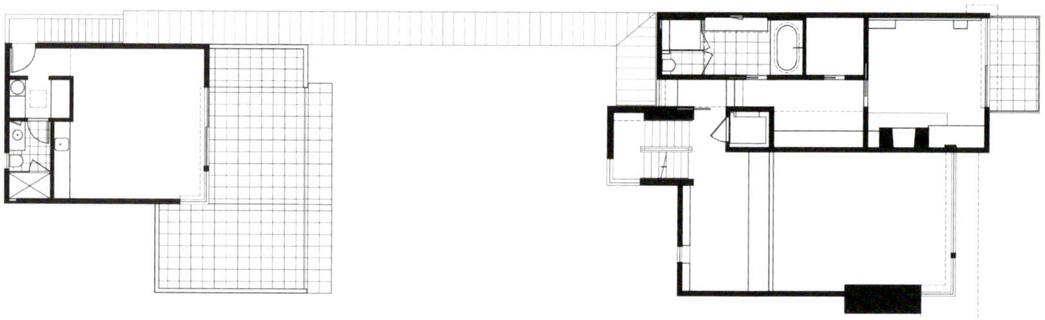

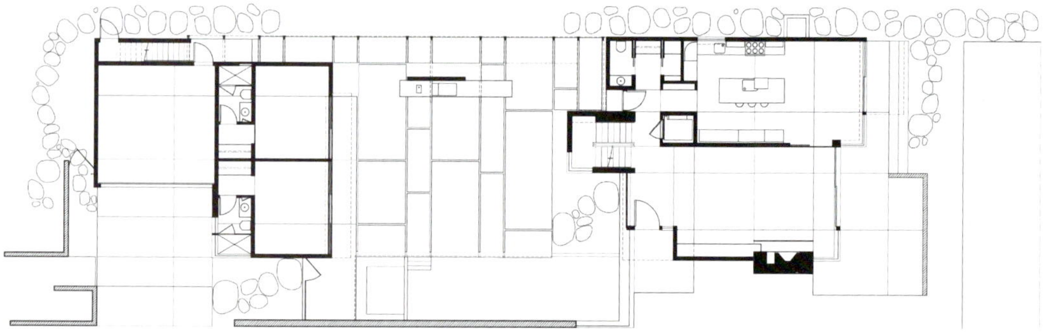

JULIUS SHULMAN
MODERNISM

VUILLARD

Wine

House IJburg

艾瑟尔堡别墅

Total gross area: 215 m² 总建筑面积：215m²

Architect: Marc Koehler Architects 建筑设计：马克 · 科勒建筑师事务所

The house is designed as a vertical garden giving space to flora and fauna to grow in a densely urbanized area. Closed private spaces contrast with open collective spaces that seem to have been "carved out" from the solid volume as a continuous transparent void. In this way the interior space is visually and physically connected to the street, the garden and roof terraces. Outdoor and indoor spaces become one and natural daylight flows into the interior.

The ornamental masonry is not only a decorative enhancement of the sculptural character of the design, but also functions as an underlayment for different sorts of climbing plants to grow up the façade, giving birth to the idea of a vertical garden, which was enhanced by integrating flowerpots on several levels in the façade. Hedra, Kiwi's grapes, apples and roses will over time overgrow the house and create a "natural curtain" around the living spaces and terraces, providing natural shading and privacy. The living area is situated on the first floor, thus the inhabitants will experience their garden as an integral part to unify nature and culture in a house.

设计者以塑造一座垂直花园的思路来设计这座别墅，在这样一个密集的城区为动植物提供一片空间。这座房屋被设计成一个单体的雕塑块状结构，与开放的集合空间形成对比。这些空间似乎是从固体中切割出来的，并连接到大街、花园和屋顶平台上。室内外空间融为一体，使室内享有良好采光。

装饰用的石材不仅为这种雕塑感锦上添花，还为各种攀爬植物创造了在立面上生长的条件，构成了垂直花园的要素，并通过整合各层的花盆加强这一设计。常春藤、新西兰葡萄、苹果树和玫瑰将在不同季节长满这座别墅，形成了天然屏障，为生活区和露台增加了阴凉和隐蔽性。把生活区设在一楼，居住者可以体会到花园的重要作用，它将自然和文化融合在一座建筑中。

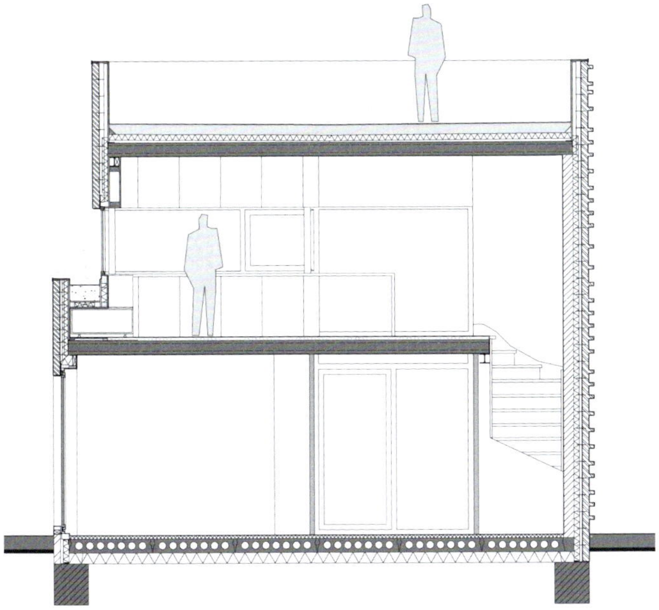

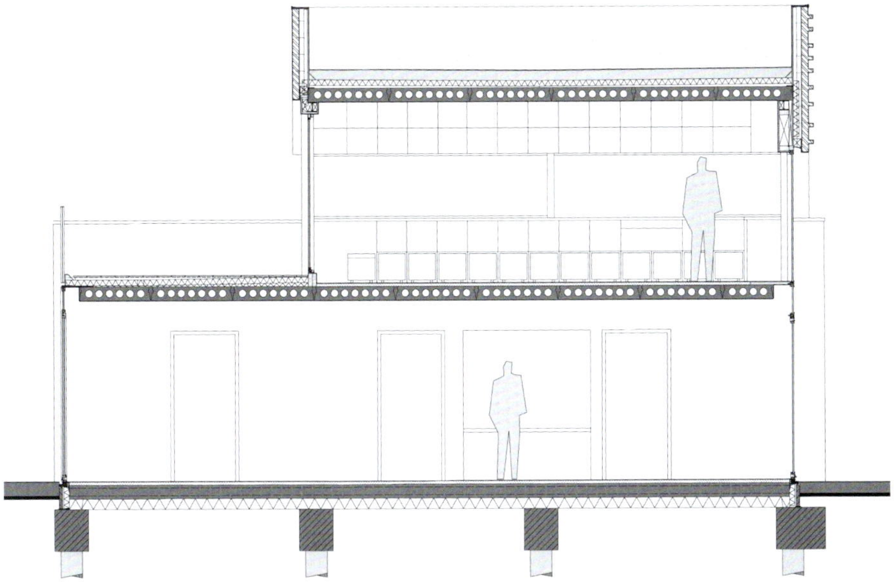

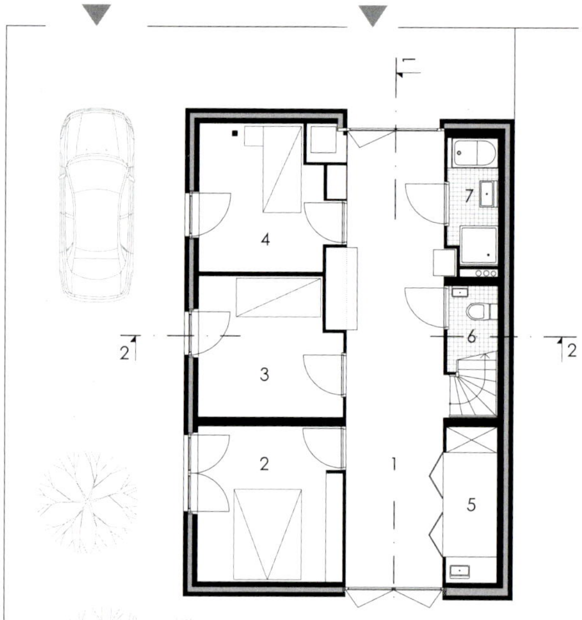

1 HALL
2 BEDROOM
3 BEDROOM
4 BEDROOM
5 STORAGE ROOM
6 TOILET
7 BATHROOM

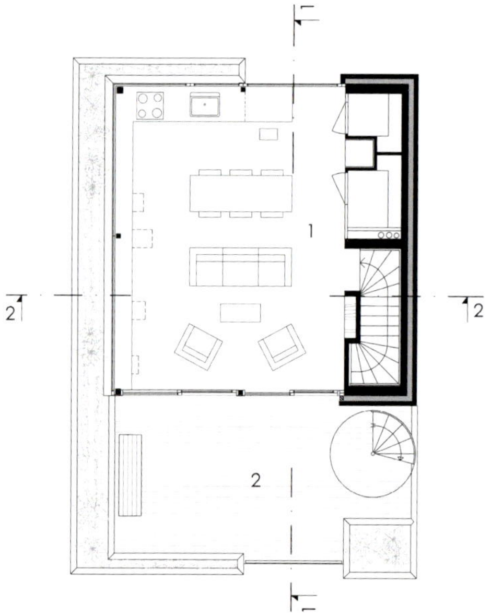

1 LIVING AREA
2 TERRACE

"Domain" Vllas

"领域"别墅

Total building coverage: 1350000 m^2

总建筑面积：1350000m^2

Designed by: Gravity Partnership Ltd.

设计公司：香港嘉柏建筑师事务所

"Domain" is located on the north bank of Min River in King-ka Town, Minhou County. It is just 15 km from the entrance of Fuzhou East Street. To its south, there is 900 m long Min River. With the Carp Isled Hotel to its opposite, the villa is adjacent to the litchi orchard of the Lai Chi Kok Resort on the west. The planed Binjiang Avenue will run across the front of the villa. The rich and natural landscape resources make the plot a rare private domain.

According to the landscape resources, on the north of "Domain", 10 short-slab style landscape apartment buildings of small high-rise will be built. They are connected with each other in curve line to form active and continuous facades along the street. This layout not only ensures the landscape and ventilation of the community, but also provides each apartment with ideal orientation and river view. On the west of the site, there are 12 semi-detached villas. Others are mainly the high-end single-family scenic villas with each covers an area of 1～2 mu.

The central lake is located in the west of the site connecting with the Min River, thus the water is running around different residential areas to create a waterfront and quiet environment. The vertical design of the community is also unique. For some residential areas without waters, the designer has lifted the base to 3 m high to provide waterscapes as well as diversify the vertical space structure of the community. Meantime, the overall layout of the community is well planned to arrange the villas in order. By adjusting the sizes of the windows, it effectively keep the privacy of the inhabitants.

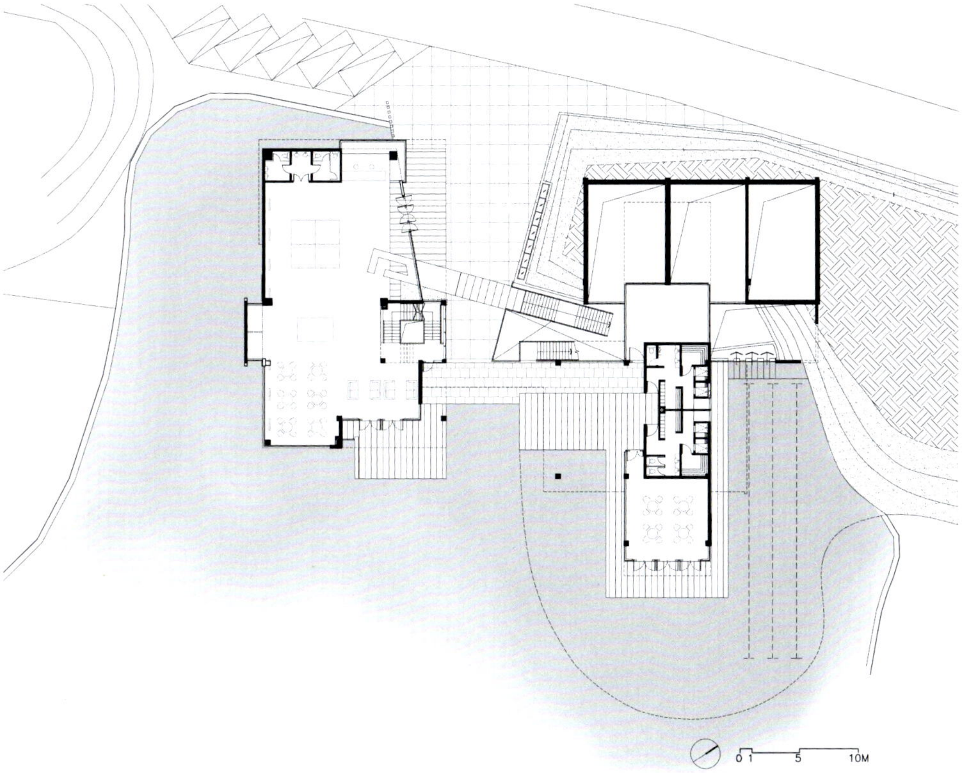

"领域"别墅项目位于福州近郊闽侯县荆溪镇闽江北岸，离福州市东街口仅15km。地块南侧紧临闽江延绵900多米，与鲤鱼洲国宾馆隔江相望，西侧与"荔园度假村"成片茂密的百年荔枝林相邻，规划中的滨江大道从项目前穿过。丰富、天然的景观资源使该地块成为不可多得的私家领域。

"领域"根据用地的景观条件，北侧布置约10幢"短板"小高层景观公寓，整体造型呈连贯的曲线，形成活泼、连续的沿街立面。既保证了小区整体的景观、通风要求，又使公寓户户取得朝向、观江景俱佳的角度。用地西侧布置了12幢双拼别墅住宅，其余大多数产品为占地1～2亩的高档独栋景观别墅。

中央内湖自然地流淌于地块的西侧，通过与闽江的活水循环，让新鲜的水体的流淌于各个组团之间，形成亲水、静谧的别墅居住环境。小区的竖向设计也比较独到，对于不可回避的不直接亲水组团，建筑师将其标高垫高3m，使其水景唾手可得的同时也丰富了小区的竖向空间形态。同时，通过社区总平的精心调配，让别墅错落布局，通过调整开窗尺度，减少别墅间的对视情况，保障业主的私密性。

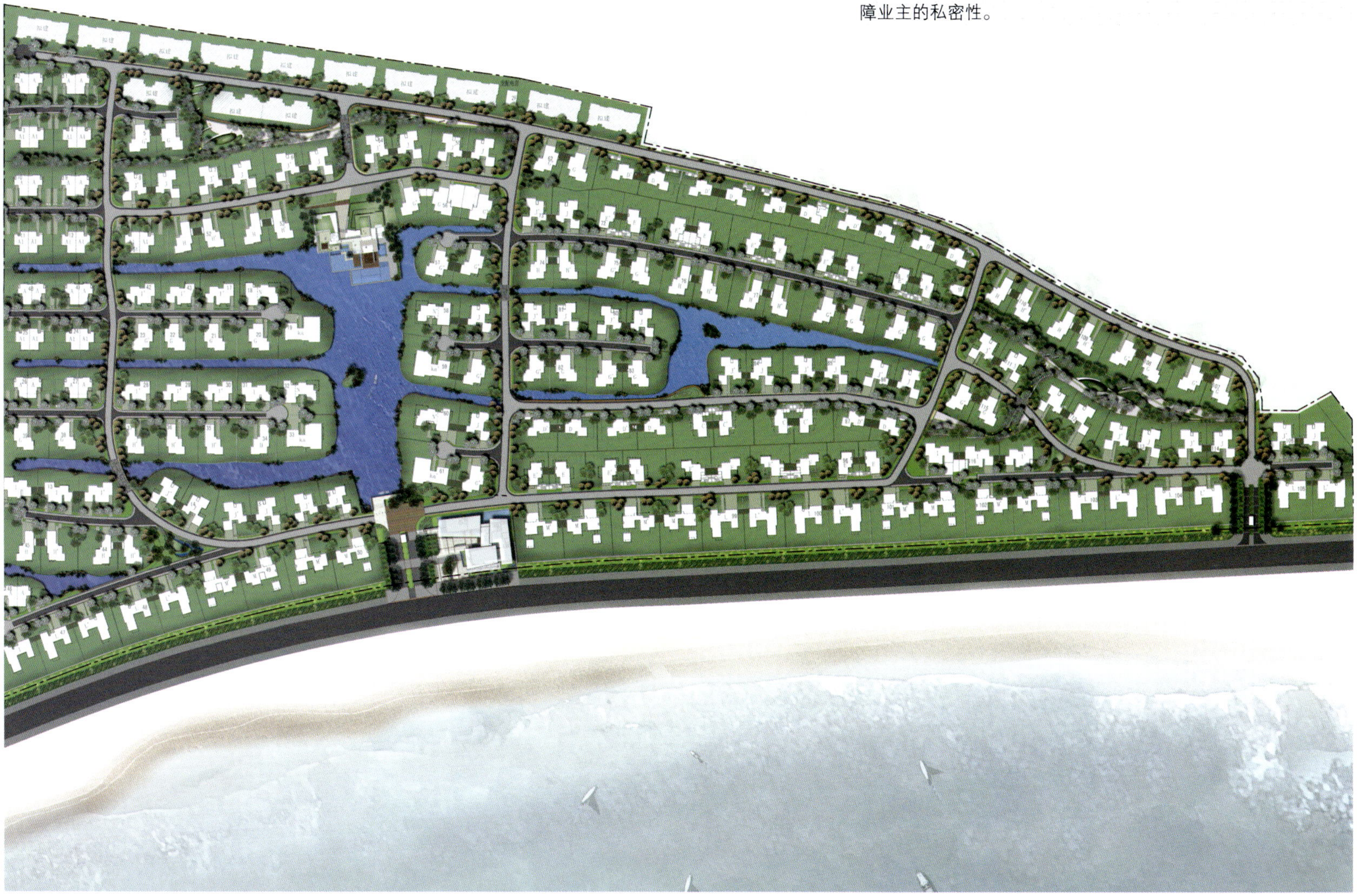

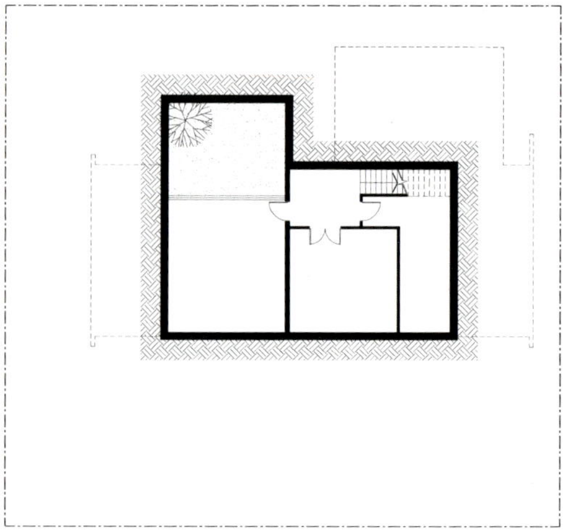

Basement Floor Plan 地下层平面图

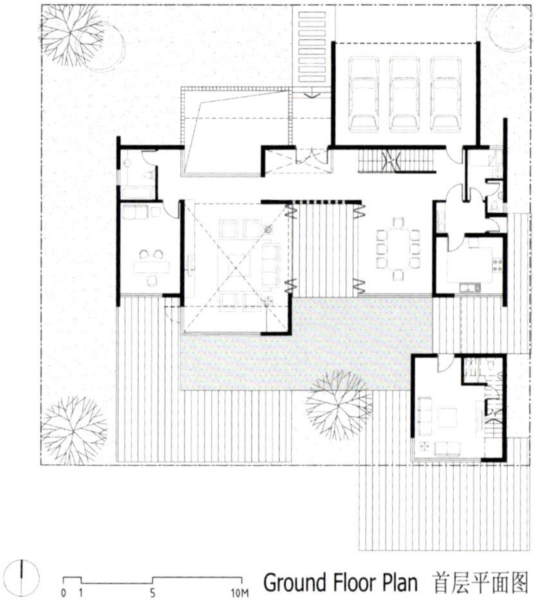

Ground Floor Plan 首层平面图

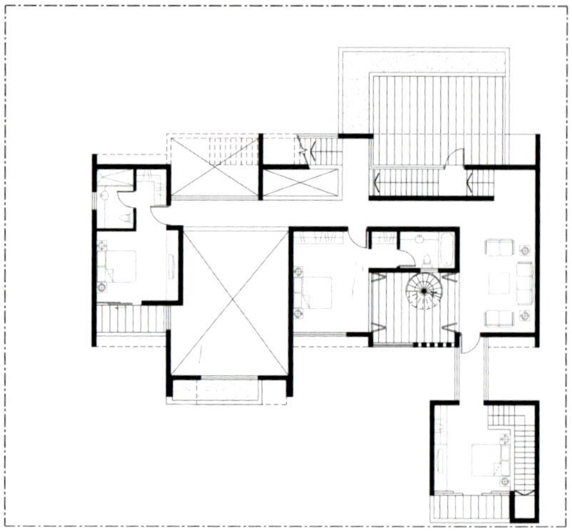

First Floor Plan 二层平面图

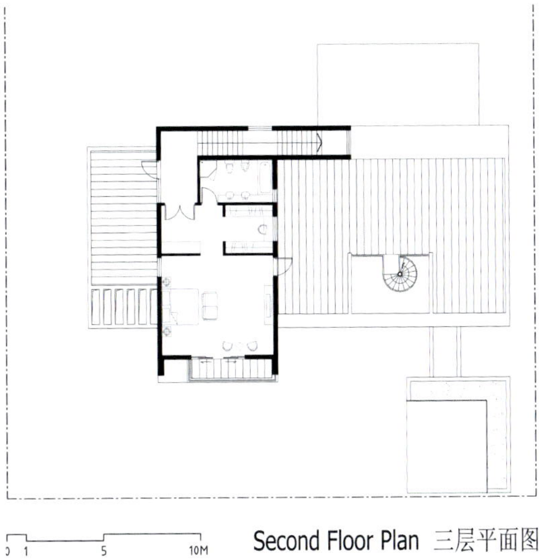

Second Floor Plan 三层平面图

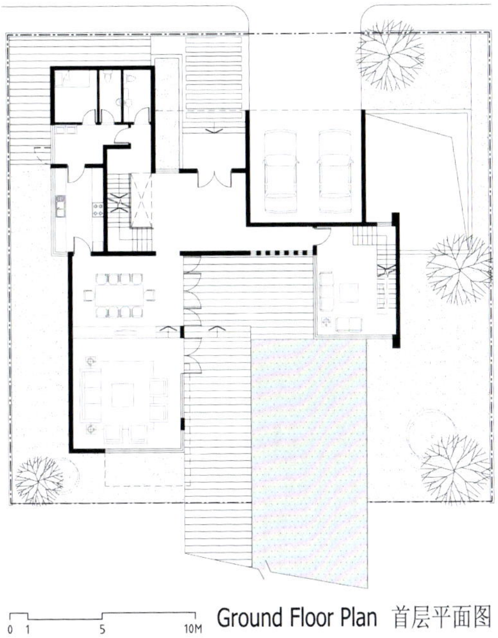

Ground Floor Plan 首层平面图

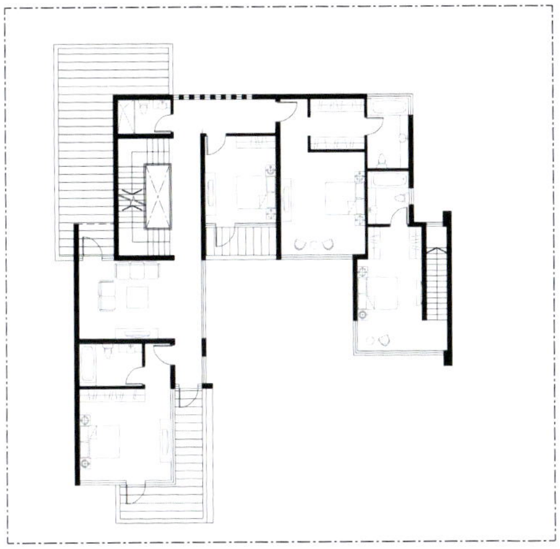

First Floor Plan 二层平面图

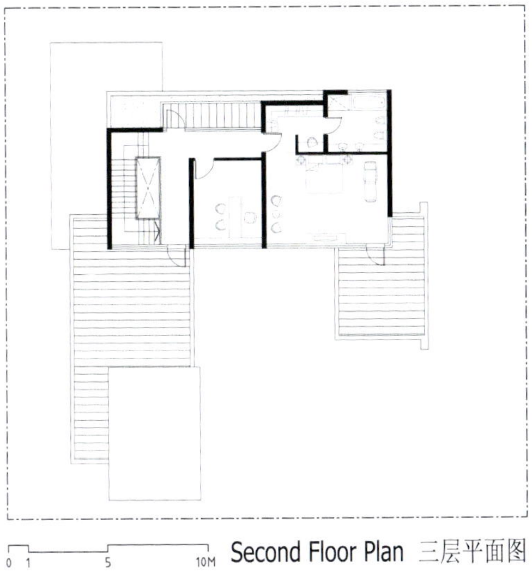

Second Floor Plan 三层平面图

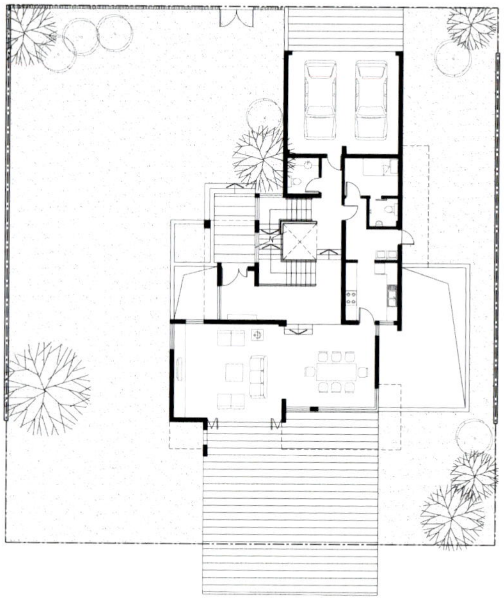

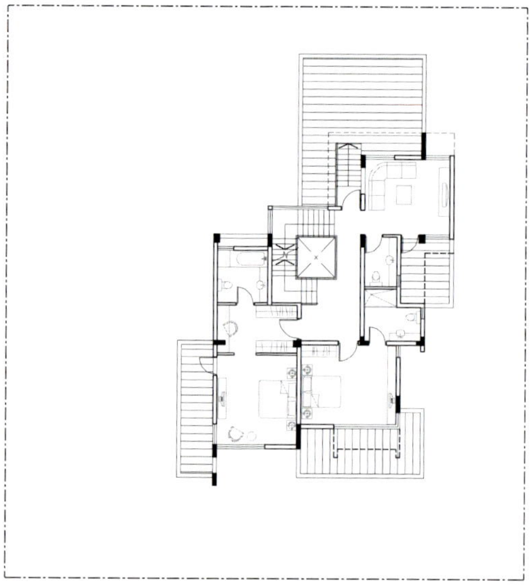

Second Floor Plan 三层平面图

Ground Floor Plan 首层平面图

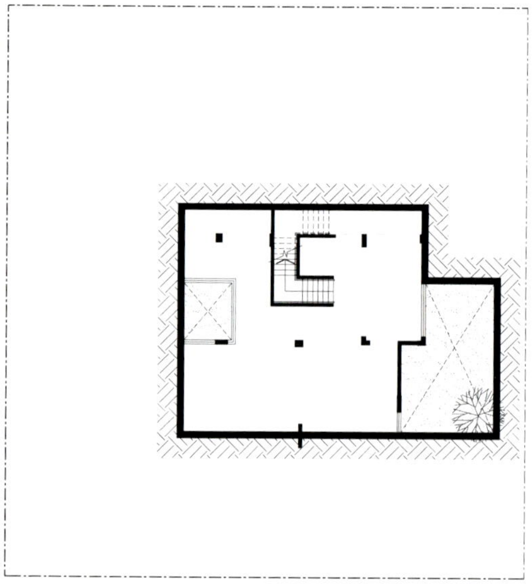

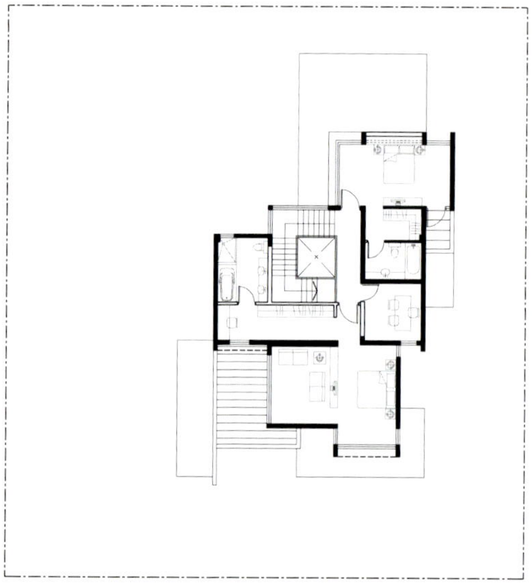

Basement Floor Plan 地下层平面图

First Floor Plan 二层平面图

Vanke • Eastern Coast Upper Residence

万科东海岸上居

Building coverage: 275000 m²　Design Company: Hong Kong Gravity Partnership Ltd.

总建筑面积：275000m²　设计公司：香港嘉柏建筑师事务所

The Upper Residence is the last part of Vanke Eastern Coast project which accumulates the essence of five-year development. Upper Residence consists of 308~420 m2 single-family villas, 252~280 m2 high-rises, 225 m2 townhouses, and 60~90 m2 boutique apartments.

The community is with fresh air and green landscapes all around. The architectures are designed in sea life style of Caribbean coast, built on the hills with massive ecological plants reserved. Walking in this community is just like walking through the forest, with changing landscapes and ecological plants everywhere.

The project makes full use of the hills and the sea resources. Dameisha is the nearest coastal beach to the center of Shenzhen. There are also paths leading to the hills back of the community. Living here, the dream of "living in a house towards the sea, with spring blossoms" will come true.

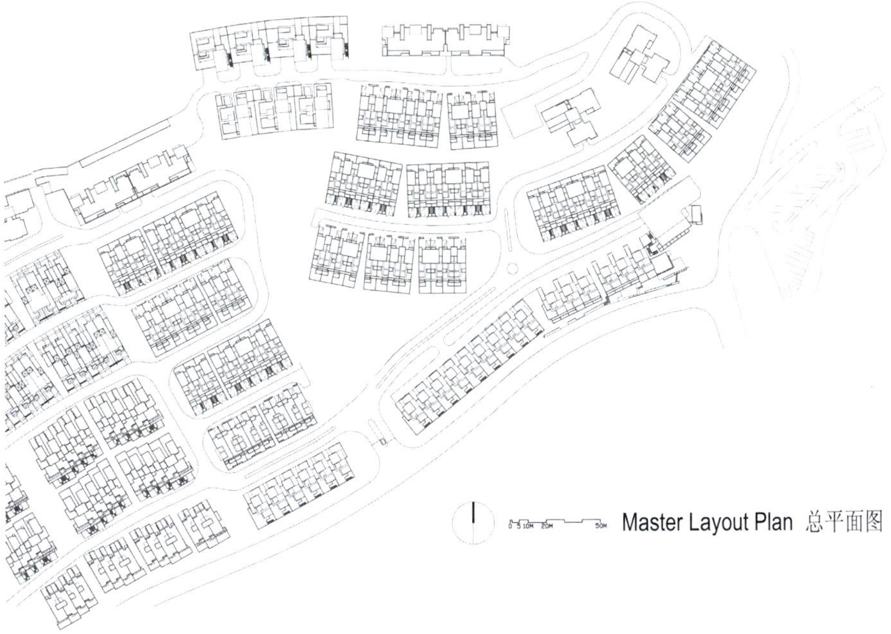

Master Layout Plan 总平面图

万科东海岸上居是万科东海岸的收官之作，积淀了项目5年的发展精粹，产品包括独栋别墅（308~420 m^2）、高层（252~280 m^2）、联排别墅（约225 m^2）和精装公寓（60~90 m^2）。

社区内部空气异常清新，满眼都是绿色。建筑采用加勒比海岸海居风格，依照山体坡势建造，大量保留了原生态植被。行走在社区的感觉就如同行走在山林中一般，景观变化丰富，原生态的植被随处可见。

项目不仅山、海兼得，而且它的山、海资源都是可以参与的。大梅沙是深圳离市中心最近的海滩，社区后面的山，也可直接进入。住在这里，“面朝大海，春暖花开”的居住梦想将得以实现。

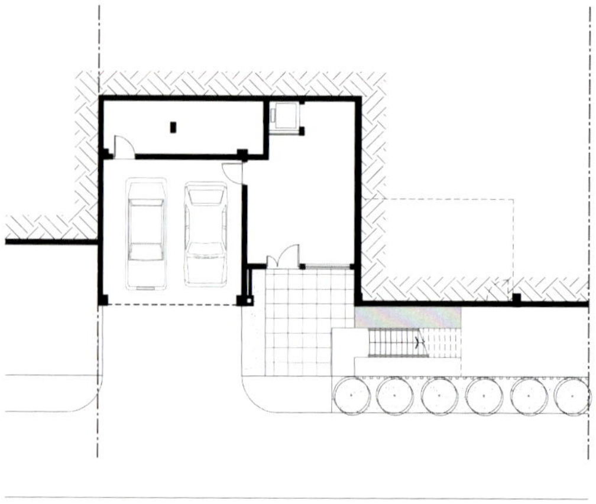
Basement Floor Plan 地下层平面图

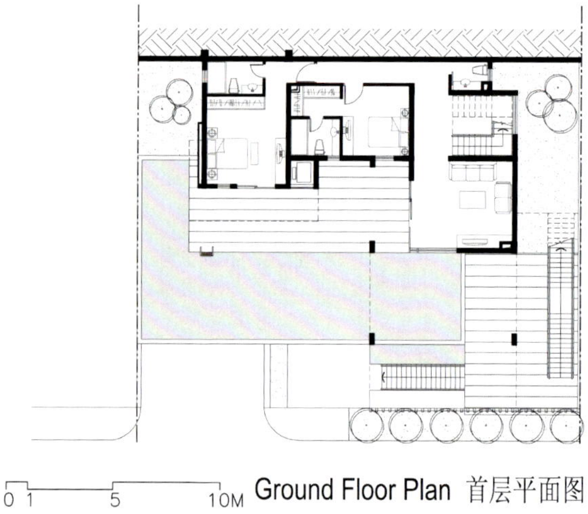

Ground Floor Plan 首层平面图

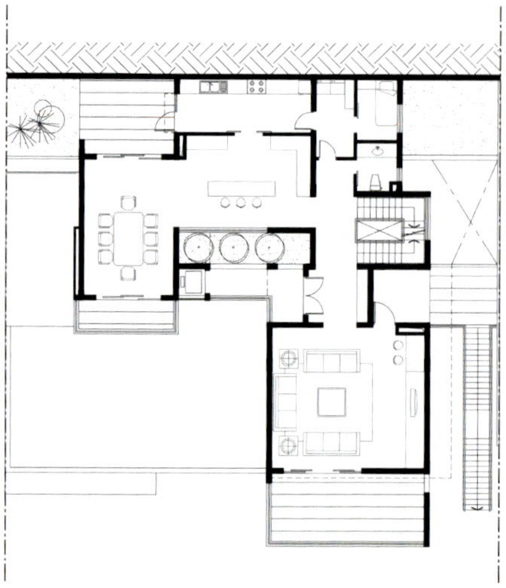
First Floor Plan 二层平面图

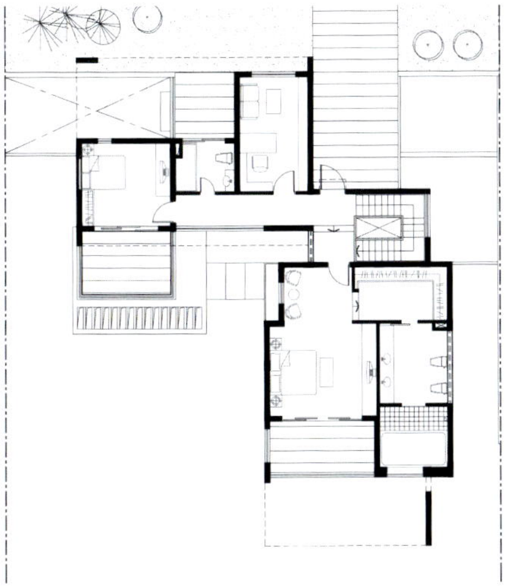

Second Floor Plan 三层平面图

Section 剖面图

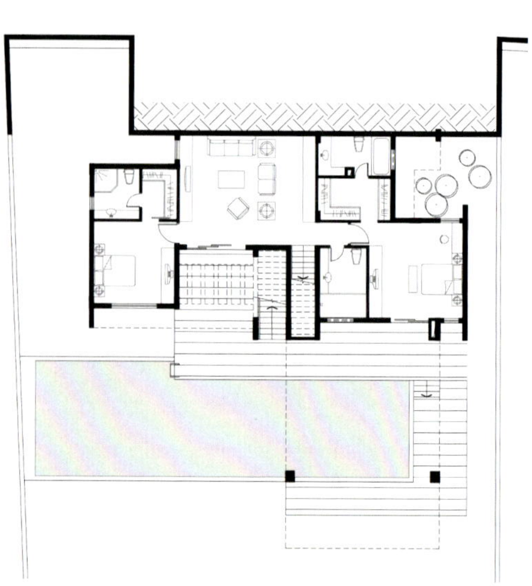

Basement Floor Plan 地下层平面图

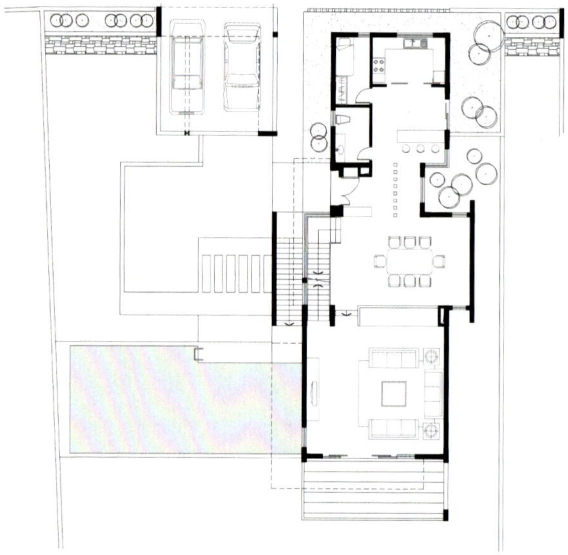

0 1 5 10M Ground Floor Plan 首层平面图

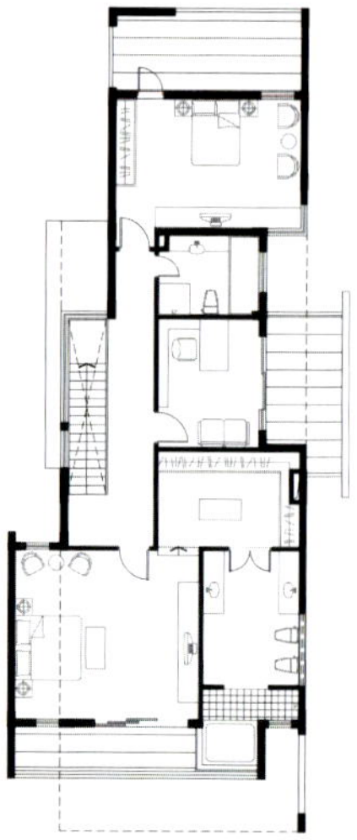

First Floor Plan 二层平面图

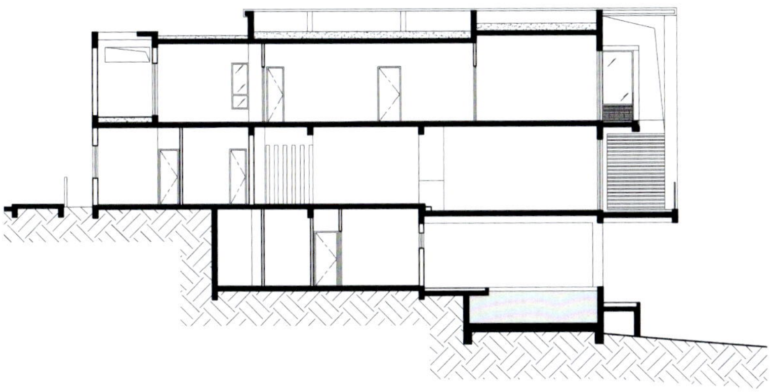

Section 剖面图

0 1 5 10M

Aegean Sea Villa

"爱琴海"别墅

Total building coverage: 94000 m^2

总建筑面积：94000m^2

Designed by: Gravity Partnership Ltd.

设计公司：香港嘉柏建筑师事务所

Aegean Sea Villa is located to the south of Xiamen Exhibition Center and to the west of island ring road. With small Jinmen Island to its opposite across the strait, Aegean Sea Villa features beautiful sea views and ideal location. A series of high-standard facilities are completed one by one, such as Xiamen International Exhibition Center, Exhibition Hotel, Garden Hotel, yacht wharf, Qianbu Park, tennis court, swimming pool, high-class club, etc. to create the atmosphere of a high-end community. Aegean Sea Villa stands in the first row of the first low-density sea-view villas and the apartments near the Island Ring Road. It is one of the top waterfront communities. The single-family villas have huge private courtyards covering more than 1000 m^2 (the biggest one is about 2700 m^2). Owning a spacious private villa in this precious land by the Island Ring Road, you can enjoy the sea view and plan in hand.

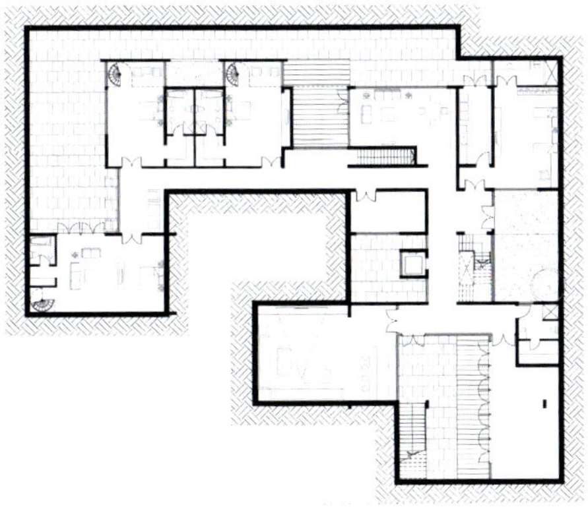

Basement Floor Plan 地下层平面图

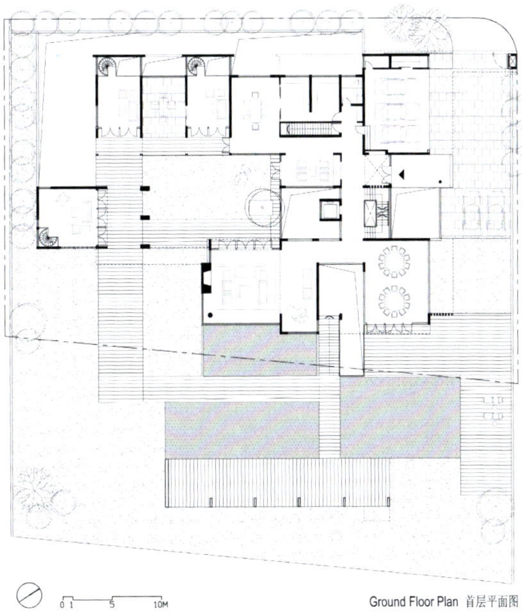

Ground Floor Plan 首层平面图

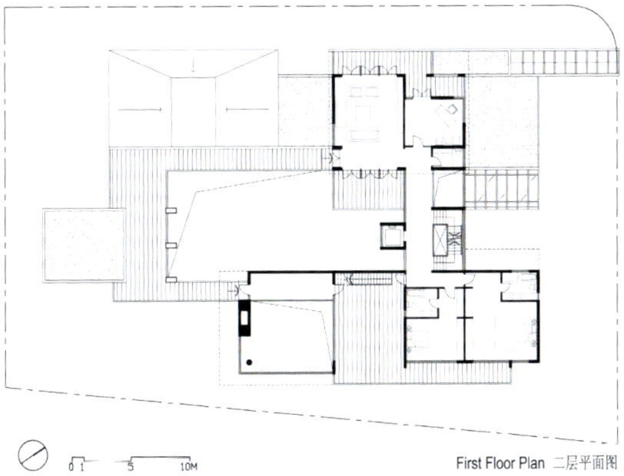

First Floor Plan 二层平面图

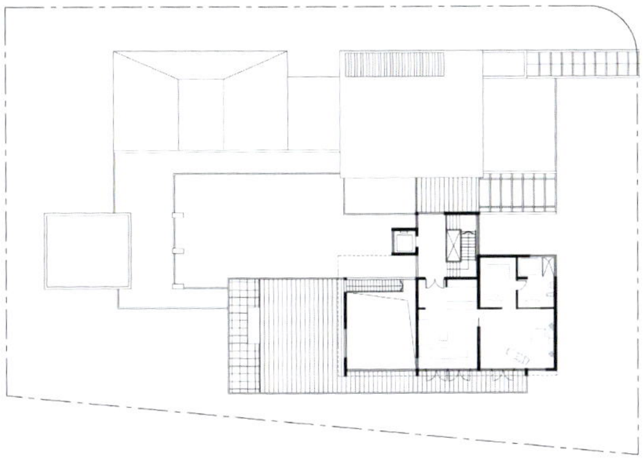

Second Floor Plan 三层平面图

“爱琴海”地处厦门会展中心南侧、环岛路以西，正对海峡对岸小金门岛，海景资源十分丰富，区位条件优越。随着厦门国际会议中心、国际会展酒店、国际花园大酒店、游艇码头及前埔公园、网球场、游泳池、高级会所等一系列高标准国际性配套的相继完善，周边高档社区的氛围逐渐形成。"爱琴海"别墅为环岛路第一排低密度顶级海景别墅及海景公寓，是最顶级的滨海豪宅社区之一，独体别墅拥有上千平米的超大私家院落（最大私家院落近2700m^2），在寸土寸金的环岛路边拥有广阔的私家领地，画海为疆，运筹帷握。

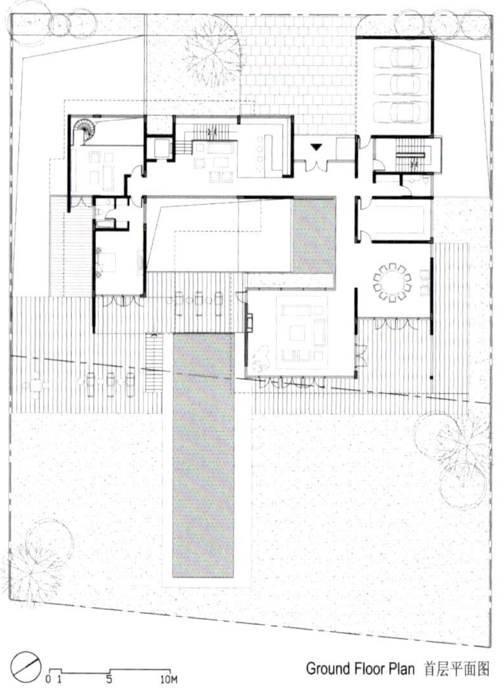

Ground Floor Plan 首层平面图

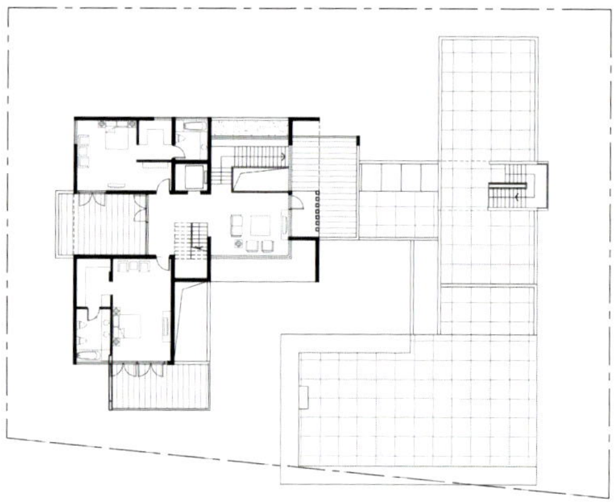

First Floor Plan 二层平面图

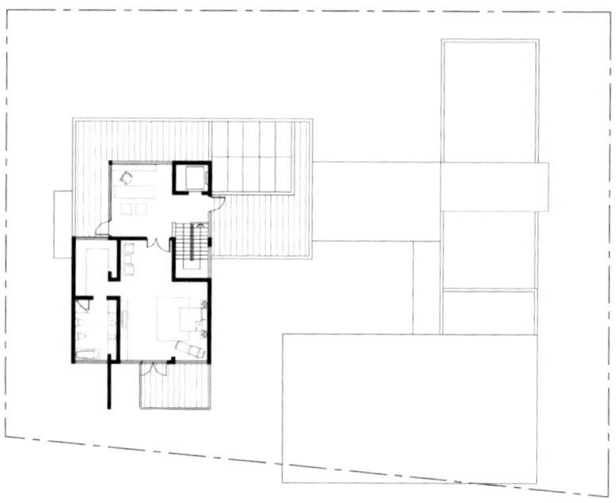

Second Floor Plan 三层平面图

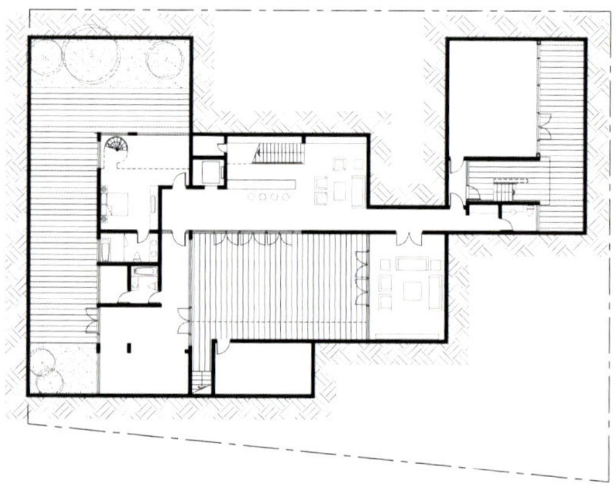

Basement Floor Plan 地下层平面图

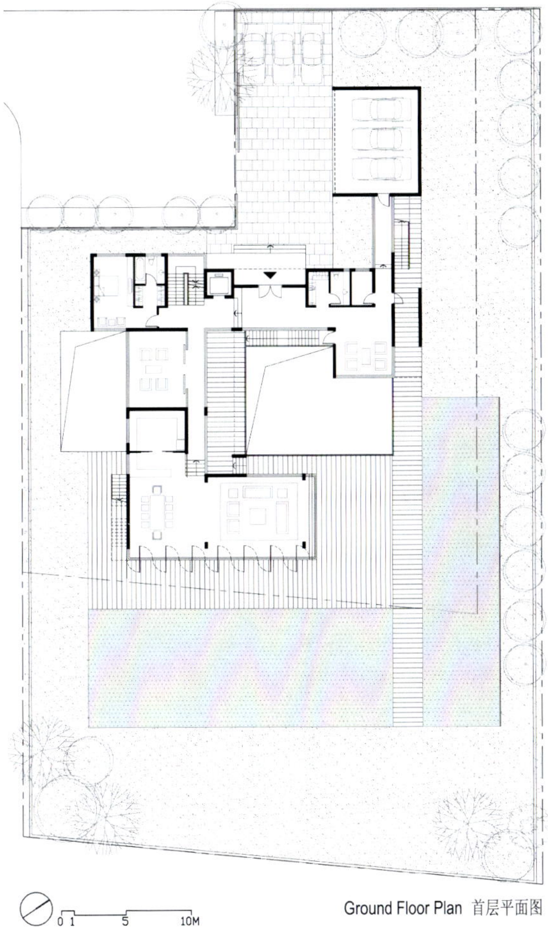

Ground Floor Plan 首层平面图

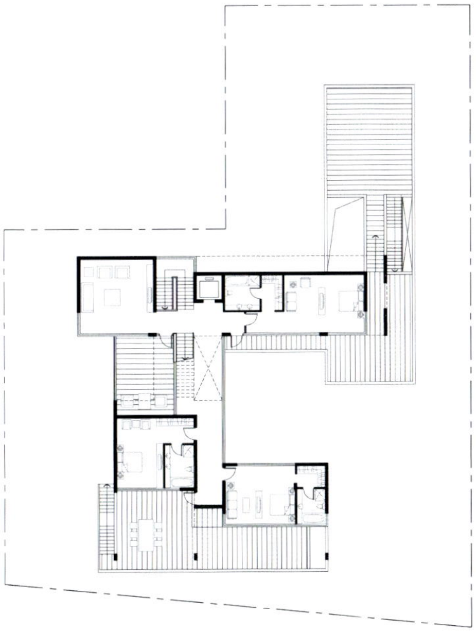

First Floor Plan 二层平面图

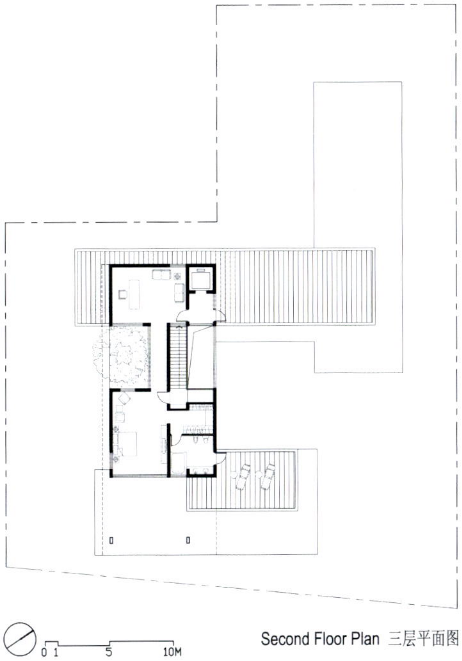

Second Floor Plan 三层平面图

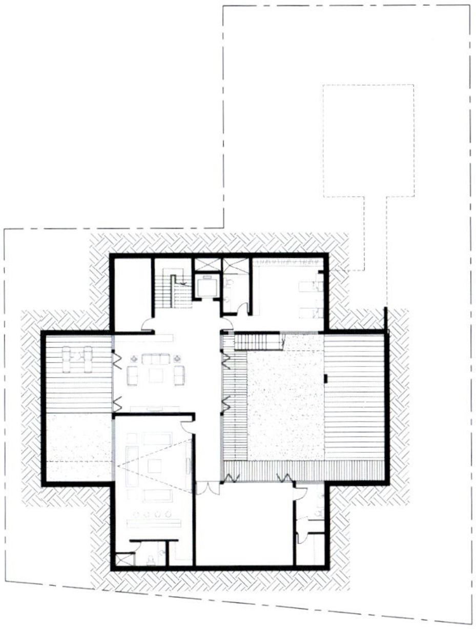

Basement Floor Plan 地下层平面图

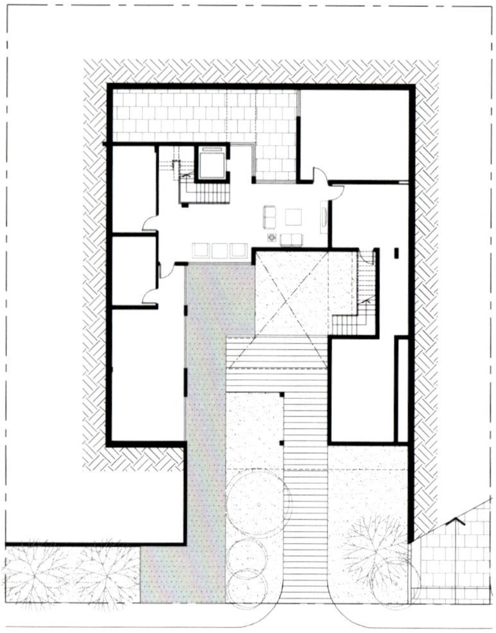

Basement Floor Plan 地下层平面图

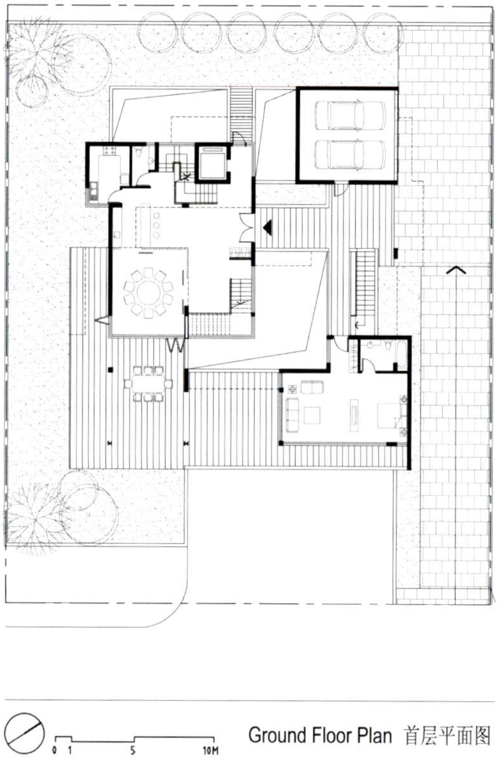

Ground Floor Plan 首层平面图

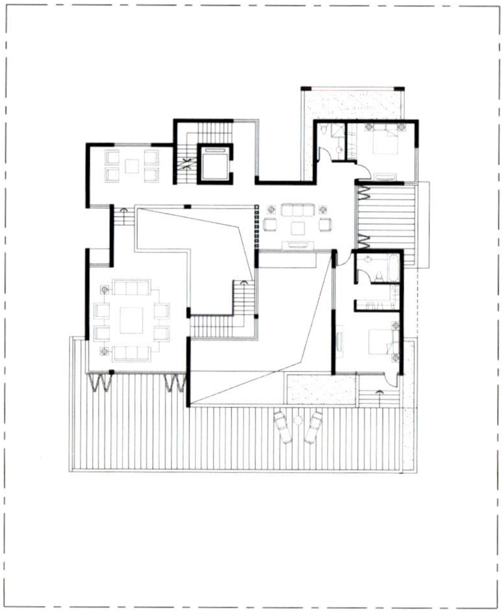

First Floor Plan 二层平面图

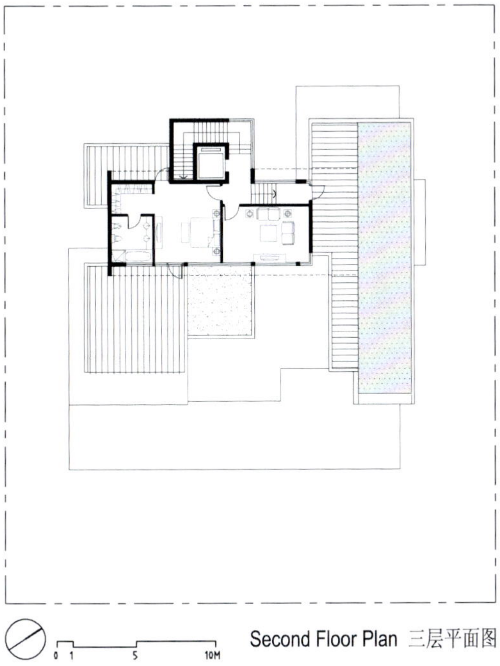

Second Floor Plan 三层平面图

House N

住宅N

Site area: 746 m²　　Designed by: 3LHD Architects

用地面积：746 m²　　设计：3LHD 建筑事务所

House N is situated in a quieter secondary neighborhood on Pantovcak street, perched in the western hills and boasting a delightful panoramic view of the forest and the city.

Directions of movement in the interior were additionally accentuated within the external volume of the house and represent its main design element. The entire house, its façades and the roof are covered with the same material, dark brown HPL panels.

The ground floor consists of a living area, which extends to the outer environment through wooden terraces, some of which are roofed over, while others are open to the sky, winding up in completely natural pebbled areas amid grass. Unlike the living room and the dining area, contact between the kitchen and the exterior is not as direct.

The main accent in the interior is a wide two floored hall that signifies the main connector of the ground and first floors. It is dominated by a hanging Poulsen "artichoke" bathed in diffuse zenithal lighting that comes from a high main window and gives the entire area a unique atmosphere.

The first floor consists of intimate areas; a work area, parent and children'

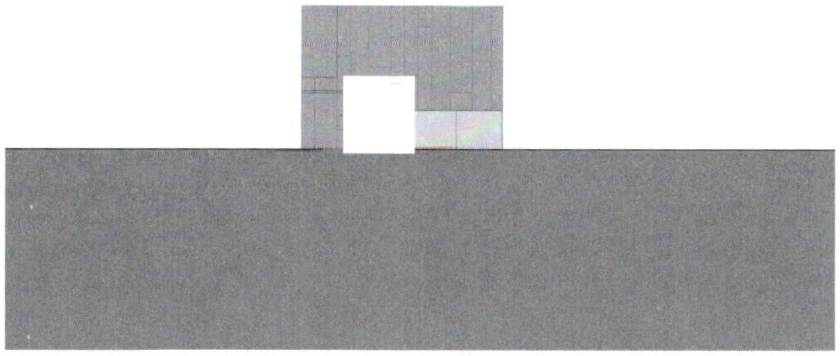

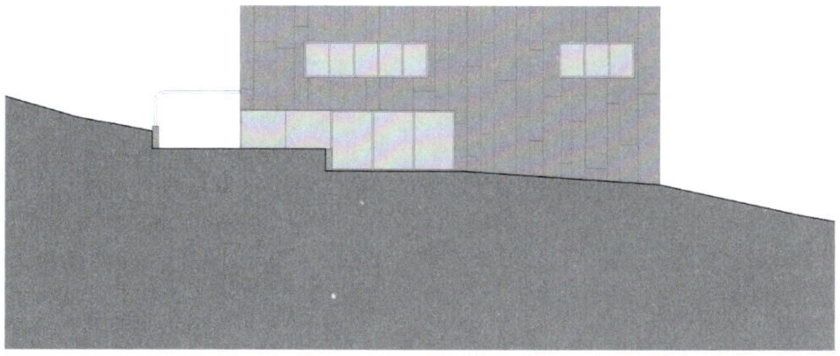

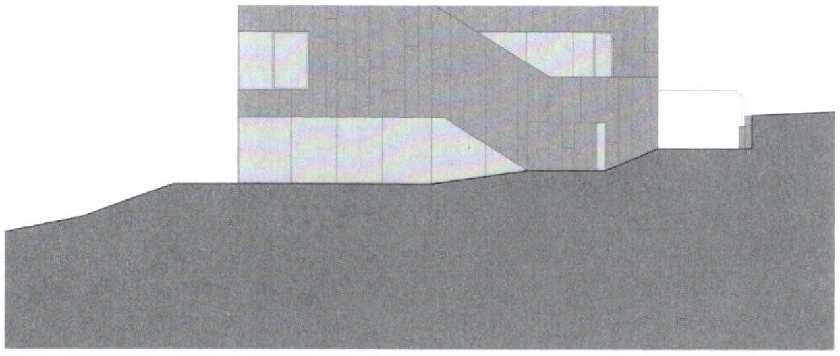

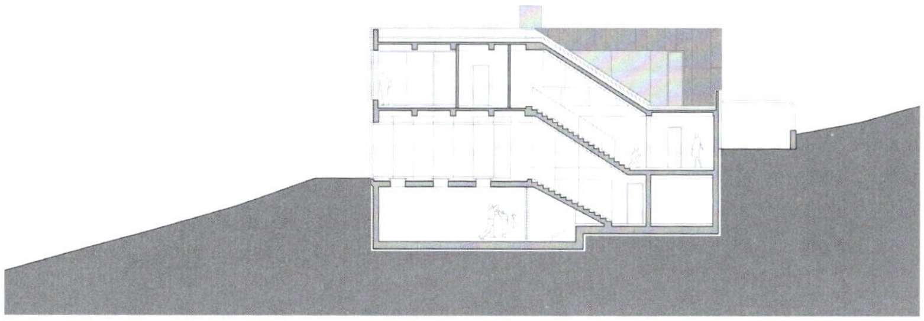

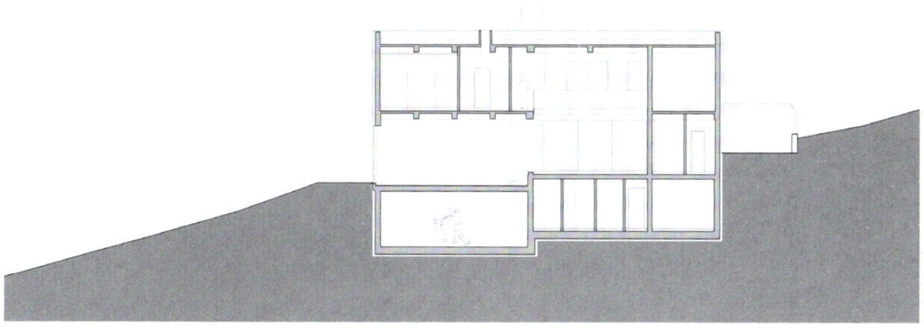

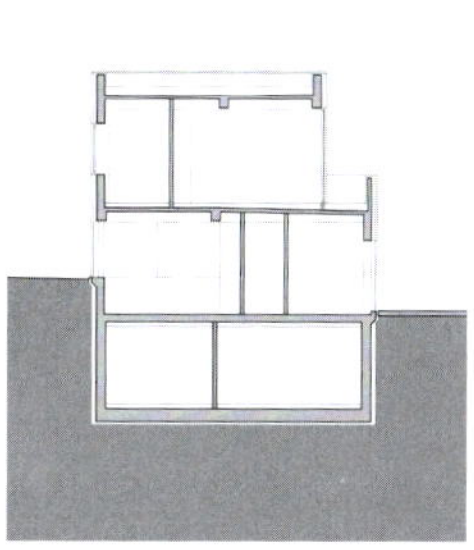

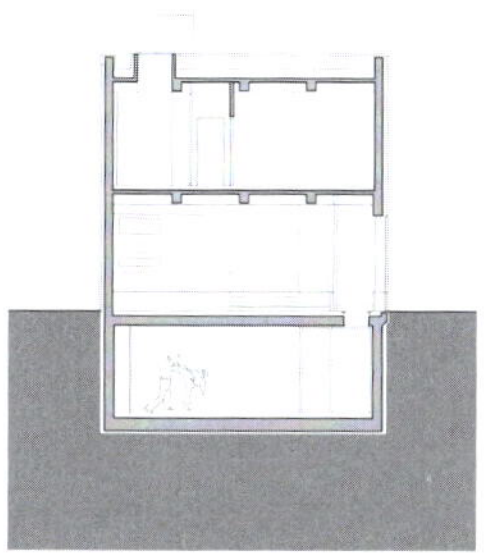

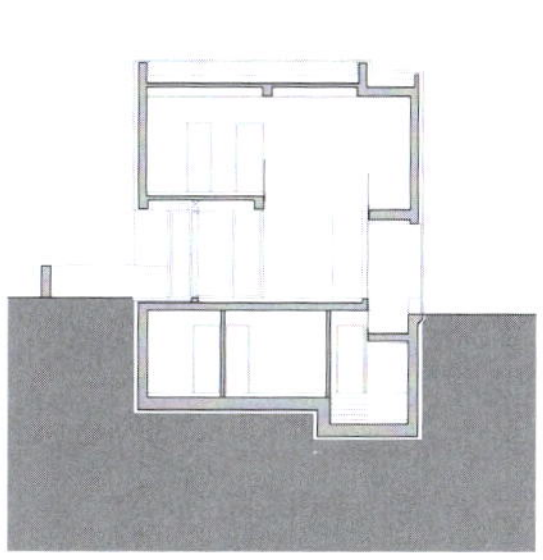

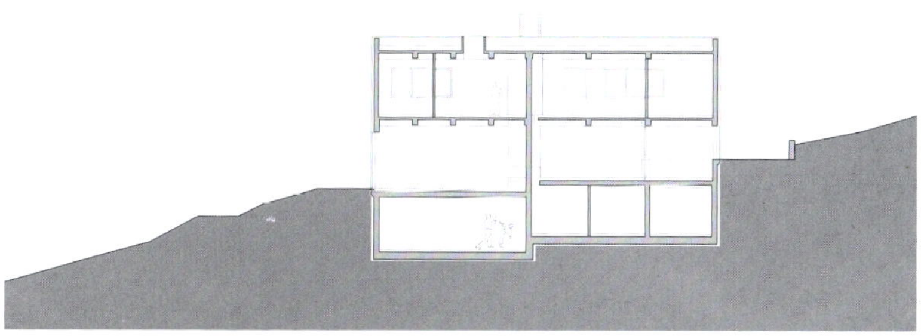

s zones, from which one may take in a most breathtaking view through a corner window.

Built-in closets on each story have been integrated completely into their surroundings with color and materials, thus enabling a minimal amount of visible and movable furniture. Floors of oiled teak and white walls create a respectable foundation for any color intervention which might take place in the future.

Special attention was given to the lighting of the rooms. Toilet and wardrobe areas are lighted with roof lights in order to provide them with as much natural and superior quality illumination as possible. The fitness room is situated in the basement and gets its only source of light through round glass ceiling lights placed along the southern glass surface of the living area floor. This position was also assigned for the placement of larger house plants in order to create an additional visual barrier towards the street. The boiler room, wellness area and wine cellar are also in the basement.

All lights are incorporated in a lowered ceiling that hides heating and air-conditioning systems, which are vented from fan coils in wardrobes and floors throughout the house. At the very bottom of the sloping lot, there is a two car garage, which is buried in the land and hidden from view of the house.

1 Entrance
2 Staircase
3 Dining room
4 Kitchen
5 Terrace
6 Storage
7 Wc
8 Living room

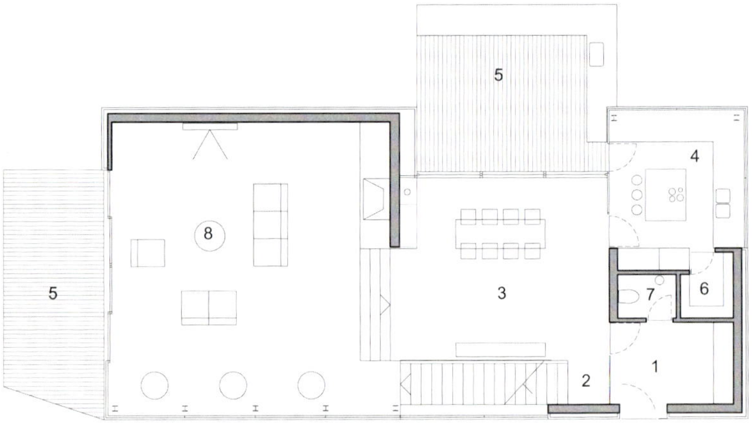

住宅N位于Pantovcak街一所比较安静的区域，西部临山，拥有壮观的森林和城市全景。

住宅的外部空间更加突出了内部的动态感，体现出它的主要设计元素。整个住宅，外墙和屋顶都采用同一材料，深棕色的高压层压板材。

第一层是生活区，由木质阳台而延伸到外面，有些阳台是带顶篷的，有些是露天的，阳台通向由纯粹天然的鹅卵石铺砌的草地。与客厅和餐厅不同，厨房不直接通向外面。

内部的主要特点就是宽敞的双层大厅，连接着一楼和二楼。上面悬挂着波尔森“菊芋”灯，沐浴在从高高的主窗户漫射的天顶采光中，整个区域独特而别致。

一楼是私密区：一个工作区，亲子区，从这里，可以通过一个角窗将壮丽景观尽收眼底。

每个楼梯的内置壁橱，在色彩和材质上与周围浑然一体，因此尽可能减少使用有形或不固定的家具。涂油柚木地板和白色墙壁营造出高尚的格调，以后可以更换为任何颜色。

特别要注意房间的采光。卫生间和换衣间采用屋顶光源照明，尽可能地使用天然或高级灯饰。健身房在地下室，唯一的照明是圆形的玻璃顶灯，装在生活区南边的玻璃墙面上。这个方向，也放置了一些大型的室内盆栽植物，专门为了隔断朝向大街的视野。锅炉房、养生区和酒窖也在地下室。

所有的灯都装饰天花板上，天花板为了隐藏供暖和空调系统而略低，供暖和空调系统是通过遍布住宅的衣柜和地板内的风机盘管进行排气。在斜坡的正底部，有一间可以容纳两辆车的车库，它位于地下，从住宅外面看不到。

9 Study
10 Bathroom
11 Wc / shower
12 Room children
13 Gallery
14 Master bedroom
15 Wardrobe

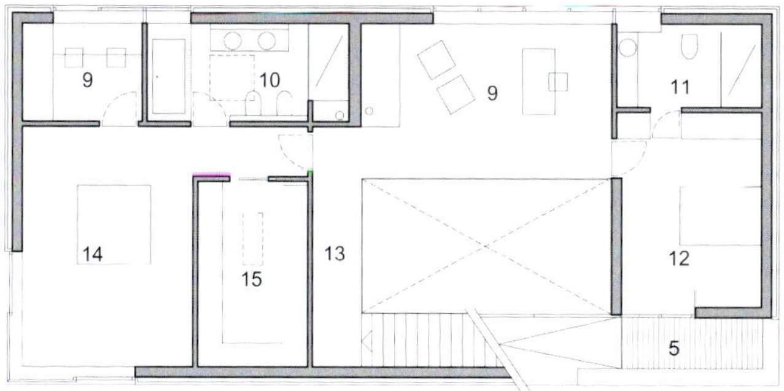

16 Fitness room
17 Boiler room
18 Storage
19 Laundry
20 Vine storage
21 Shower
22 Sauna
23 Hall

Lakefront Residence, Suzhou

苏州水云居

Construction area: 129245 m^2

建筑面积：129245 m^2

Architectural designed by: Allied Architects International (AAI)

建筑设计：加拿大AAI建筑设计事务所

To meet people's requirements for waterscapes and outing in spring, Waterfront Villa Residence connects the central green land with the waterfront platform by a green corridor. Thus all the landscapes connect with each other closely, open or closed with corridor or shading, to provide people with different feelings. The design integrates people, architectures, water and green land together to form multi-layer landscapes with the lake in the center, waters running and green walkways between.

The entrance hall is spacious and bright with two corridors leading to the living room which also connect with the garden through the outdoor corridor to create a diversified space for the inhabitants. All the master bedrooms are designed with a study room, a locker room, a washroom, a terrace and so on. The personalized workroom and recreation room connect with the sunken garden which connects with the outdoor main courtyard directly. The flexible space is well-arranged to achieve the natural transition between the public space, semi-open space and the private space.

水岸公馆为实现居住者观水与踏青的双重享受，将组团的中心绿地用一条绿色走廊与岸边亲水平台相接，环境景观环环相扣，有开阔、有封闭，有通廊、有遮挡，景致不同而耐人寻味。将人、建筑、水、绿地融为一体，形成以湖面为中心、河流相伴随、绿化步行系统相衔接的多重景观视觉，体现了对居住者景观视线的关怀。

别墅入户的门庭宽敞明亮，两层通高的通廊直接通向起居室，越过外廊与花园相接，为居住者创造了丰富的空间意境。每户主卧室均有书房、更衣间、盥洗室、露台等服务空间，富有个人情趣的工作室、娱乐室与下沉庭院相通，可直达户外主庭院。灵动空间分区动静合理，体现从公共区域、半公共半私密空间到私密空间的自然过渡。

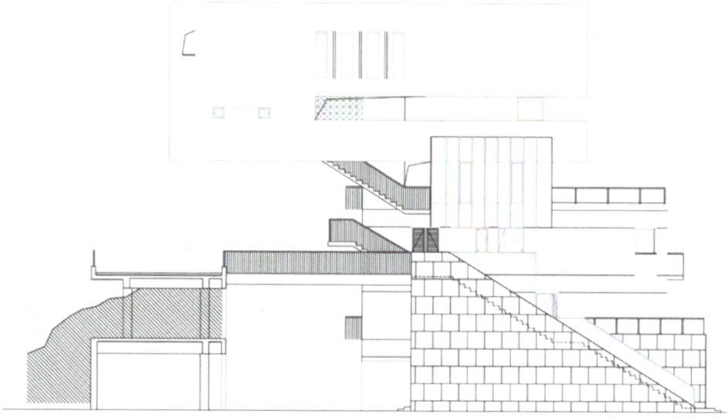

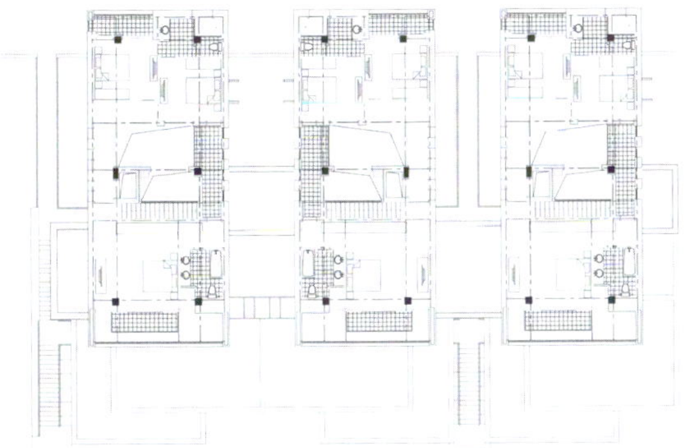
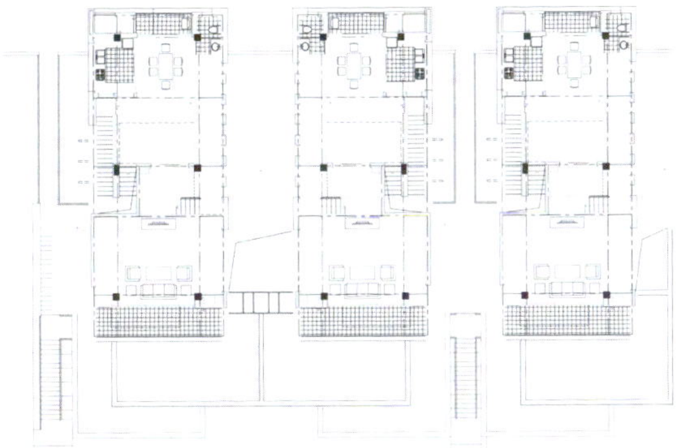
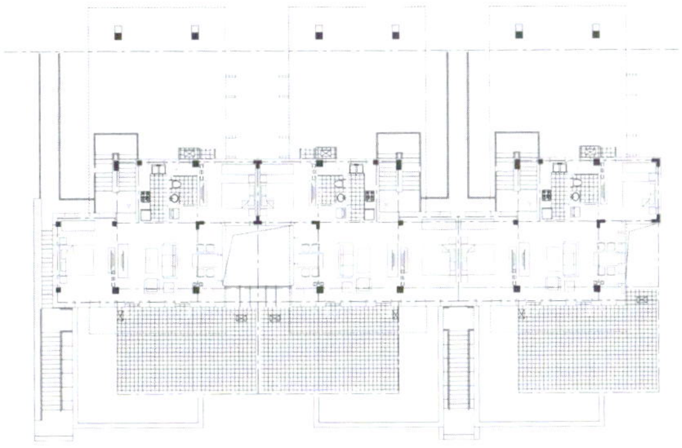

Phase I of The Bay, Shanghai

上海青浦区涵璧湾花园一期

Construction area: 194959 m^2 Architectural designed by: Atelier FCJZ

建筑面积：194959 m^2 建筑设计：非常建筑

The Bay is located in Qingpu District which is positioned to be a holiday villa. The site was used to be the fishpond which owns 430,000m^2 pond. In the breeding season, a large number of water birds rest here because of the good ecological environment. The Bay is on Island B which is located in the north part of the site, consisting of 20 buildings with 5 house styles. The floor areas of the buildings range from 514m^2 to 1022m^2.

The design tries to integrate the architectures with the natural environment as well as the cultural environment. Water is one of the important elements for the natural environment; traditional Jiangnan architecture is the mainline of the cultural environment. The modern lifestyle and the construction conditions determine that the design of The Bay cannot be the simple repetition of the tradition.

The design focuses on three key words: "division", "courtyard" and "garden".

Division: the design divides the whole project into several parts according to their different functions. The project has become a collection of a series of single buildings. Thus more rooms can enjoy better ventilation and daylight to fit the local climate. Meantime, the interior spaces can connect with the outdoor spaces and landscapes closely.

Courtyard: the buildings are divided and recombined to form many closed and semi-open courtyards which provide outdoor spaces for the inhabitants.

Garden: the walkways and waterscapes lead people to a kind of recreational lifestyle which echoes the traditional Jiangnan Garden. Thus every villa is not only a house, but also a micro-garden. People can live here and also travel here.

For the form, the gable wall and the slope roof are the architectural elements of Jiangnan dwellings. And the utilization of modern materials, such as the grey stones, aluminum alloy doors and windows, and the metal capping, has reinterpreting the tradition.

涵璧湾花园地处上海青浦区，定位是度假别墅，用地原为养鱼塘，拥有近43hm²的天然湖面，在繁殖季节，有大量水鸟在此栖息，生态环境良好；涵璧湾地块位于园区北侧B岛，一共20栋，5种户型，地上规模514～1022m²不等。

设计希望建筑和环境融为一体。这个环境既是自然的也是人文的：水是当地自然环境中的要素，江南建筑传统是这里人文环境的主线。 同时当代的生活方式和建造条件也注定建筑不可能是传统的简单重复。

设计围绕三个关键词展开——“分”、“院”、“园”。

分：化整为零，将建筑内各功能进行拆分重组，使一个建筑更像若干个建筑的集合；由此，可以使更多的房间前后通透，得到良好的通风、采光质量，适应当地阴湿多雨的气候， 同时与室外空间与景观更紧密地咬合。

院：拆分重组后的房与房之间，形成多个不同尺度的围合、半围合院，提供给住户可居的室外空间；

园：从路到水的景观系列引人进入休闲生活的状态，也呼应了传统江南园林的体验；至此，每个别墅既是一个房，还是一个微缩的园；对它的使用既是住，也是游。

形态上，山墙加坡屋顶再次运用江南民居的建筑元素，而非传统材料做法——灰色石材、铝合金门窗、铝合金屋面、金属压顶——等的运用，又是对其的重新诠释。

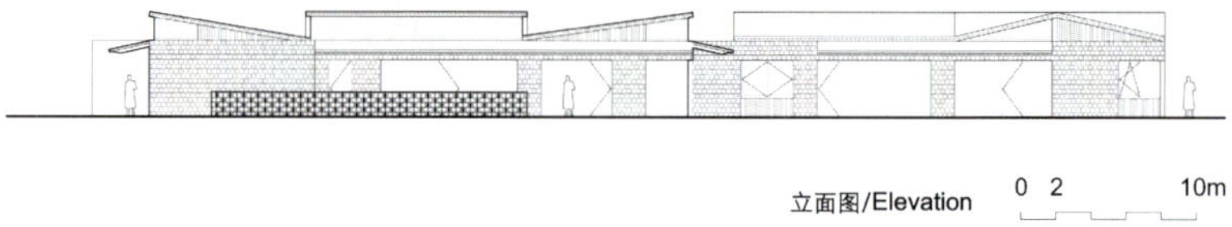

立面图/Elevation

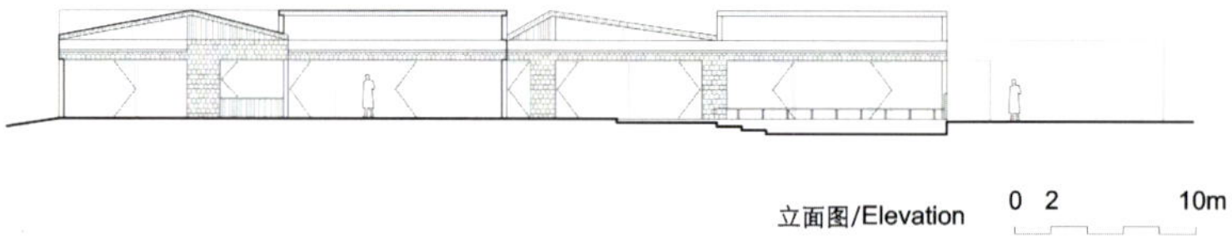

立面图/Elevation

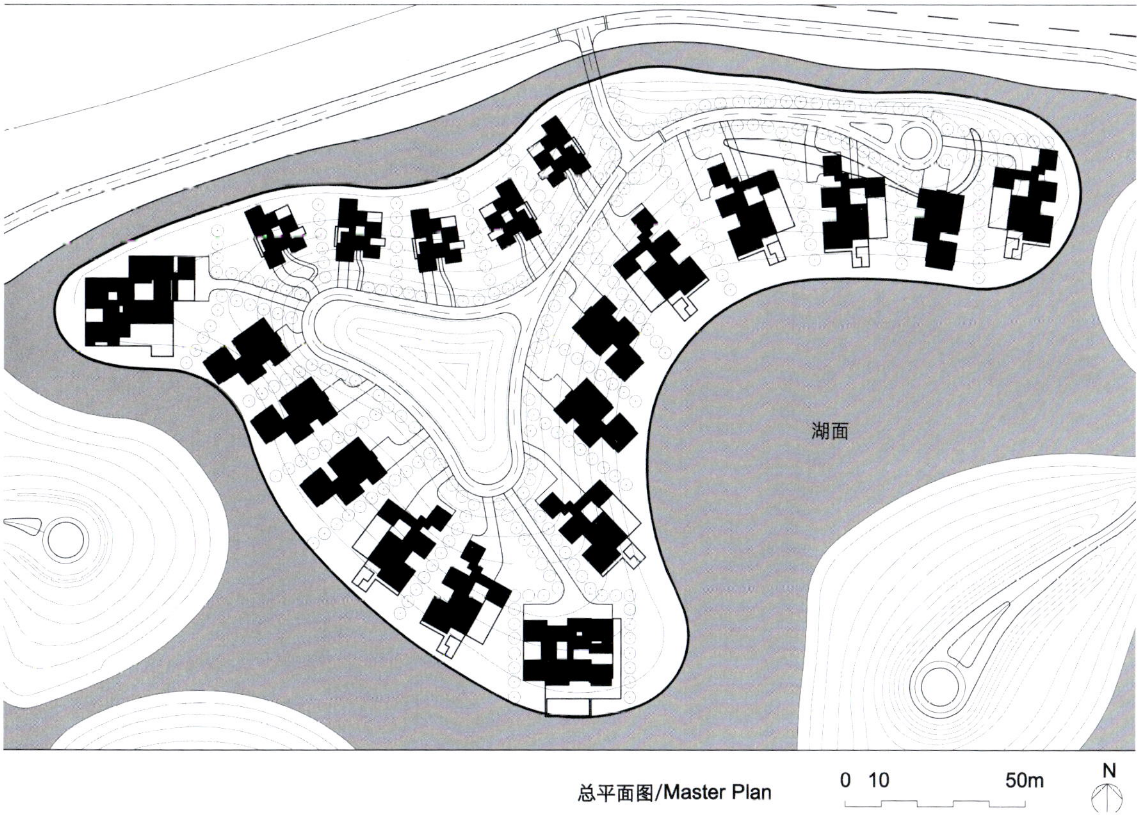

总平面图/Master Plan

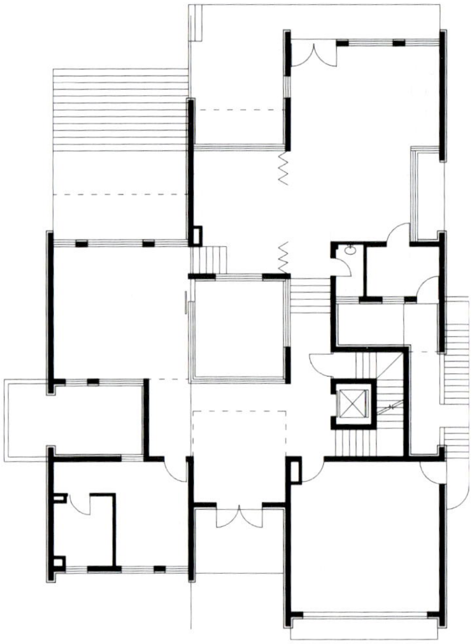

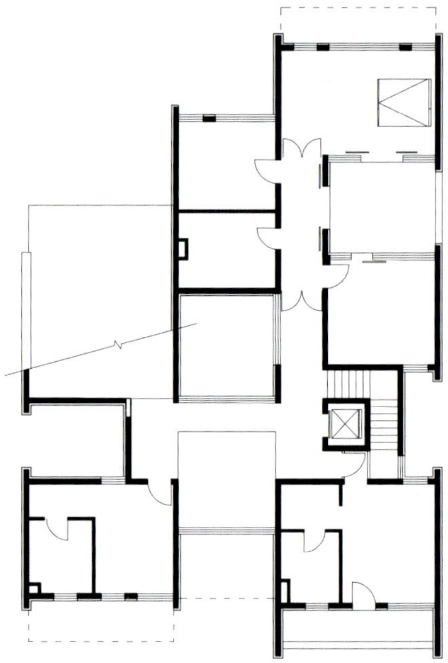